A SOCIAL CAPITAL INTEGRATED APPROACH TO URBAN CONSERVATION OF HISTORIC QUARTERS: A STUDY OF TWO CHINESE CITIES

Han Jie

中国建筑工业出版社
CHINA ARCHITECTURE & BUILDING PRESS

图书在版编目（CIP）数据

历史街区保护的社会资本整合方法：以两处中国城市历史街区为例=A SOCIAL CAPITAL INTEGRATED APPROACH TO URBAN CONSERVATION OF HISTORIC QUARTERS: A STUDY OF TWO CHINESE CITIES：英文/韩洁著. —北京：中国建筑工业出版社，2019.7

ISBN 978-7-112-24053-1

Ⅰ.①历… Ⅱ.①韩… Ⅲ.①城市道路—城市规划—研究—中国—英文 Ⅳ.①TU984.191

中国版本图书馆CIP数据核字（2019）第156642号

责任编辑：率 琦
责任校对：党 蕾

A SOCIAL CAPITAL INTEGRATED APPROACH TO URBAN CONSERVATION OF HISTORIC QUARTERS: A STUDY OF TWO CHINESE CITIES

Han Jie

*

中国建筑工业出版社出版、发行（北京海淀三里河路9号）
各地新华书店、建筑书店经销
北京点击世代文化传媒有限公司制版
北京中科印刷有限公司印刷

*

开本：787×1092毫米 1/16 印张：17¾ 字数：442千字
2019年7月第一版 2019年7月第一次印刷
定价：75.00元
ISBN 978-7-112-24053-1
（34531）

版权所有 翻印必究
如有印装质量问题，可寄本社退换
（邮政编码 100037）

Acknowledgement

The past few years in Singapore have been a journey to rediscover myself. Though it is true that PhD is a longy journey, it is also an amazing one for me to have received so many blessings. I am greatly indebted to a lot of people. First and foremost, I'd like to express my deep and sincere gratitude to my supervisor, Professor Heng Chye Kiang for his inspiring guidance, constructive suggestions, enormous patience, constant trust, as well as invaluable support and encouragement during the entire process of my study. I am highly honored and blessed to be his student, which will always be an enlightening and memorable experience in my life.

I am also grateful to Professor Wang Qiheng from Tianjin University, Professor Nezar Alsayyad from U.C Berkeley, and Professor Tang Yali from Xi'an University of Architecture and Technology for their kind support, invaluable comments and constructive advice at several critical moments of my study.

Special thanks go to Dr. Huang Zhipeng from School of Science and two members of CASA, Dr. Yeo Su-Jan and Dr. Chamari Edirisinghe who encouraged me and helped me a lot in statistical analysis and editing this piece of work; I would like to extend my appreciation to other members of CASA, as well as to the family of Ms. Bai Lili and Mr. Yang Fan, and of Ms. Xue Yuhong and Mr. Xia Lei, and of Dr. Li Suyu for their generous support and kind help.

I am sincerely grateful to the leading and guidance of my mentors Ms. Chew Moh Leen and Dr. Chuang Shaw Choon, who have been constantly encouraging me during the past years; and to my brothers and sisters for being family. Special thanks go to my friends: Dr. Wang Yan, and Ms. Jiang Xiaoqin for great friendship which have been encouraging me during this journey.

Last but not the least; I owe my deepest gratitude to my beloved parents and husband for their patience, faith, sacrifice, and unconditional love that empower me throughout my study. They are angels of great gentleness, kindness, patience, and courage, without the sacrifice of my family and their unconditional supports, I could not imagine that this piece of work would even be achieved.

Han Jie
June 16, 2019, Xiamen

Table of Contents

Acknowledgement
List of Figures
List of Tables
List of Charts

Chapter 1　Introduction ··001
    1.1　Significance of urban conservation in urban development ············001
    1.2　Research background and context ···004
        1.2.1　Research background——urban conservation and sustainable development ···004
        1.2.2　Research context——Chinese cities ·······································009
    1.3　Statement of the research topic ··011
        1.3.1　Identification of the knowledge gaps ·····································011
        1.3.2　Research questions ···012
        1.3.3　Research hypotheses ··013
        1.3.4　Research objectives ··013
    1.4　Potential contribution of this research ··014
    1.5　Scope and scale of the study ··014
        1.5.1　Scope of the study ··014
        1.5.2　Scale of the study ···016
    1.6　Outlines of the book ··016

Chapter 2　Literature Review of Urban Conservation ····················018
    2.1　Definition of urban conservation ···018
        2.1.1　Urban conservation, preservation and restoration ··················018
        2.1.2　Heritage and heritage value ··021
    2.2　The evolution of urban conservation doctrine in Europe (1800s—2000s) ···023
        2.2.1　The emergence of modern "Architectural Conservation" (1800s—1930s) ··023
        2.2.2　The rise of "area-based conservation planning" and the struggles to balance conservation and development (1940s—1960s) ·····················024

  2.2.3 The "policy—initiated Conservation" (1970s—) ········ 026
  2.2.4 New agendas of urban conservation (1980s—2000s) ········ 028
 2.3 Urban conservation practices in global perspective
   (1970s—2000s) ········ 029
  2.3.1 Approaches adopted for urban conservation in historic areas
    in Europe and the United States ········ 029
  2.3.2 Approaches adopted for urban conservation of historic areas
    in Singapore and Japan ········ 033
 2.4 Urban conservation in China ········ 038
  2.4.1 The history of urban conservation in China ········ 038
  2.4.2 Current conservation strategies of urban historic quarters in China ········ 040
 2.5 A summary of current research approaches on
   urban conservation ········ 043
  2.5.1 Approaches in the cultural and historical dimensions
    for historical sites ········ 043
  2.5.2 Approaches in economic dimension ········ 045
  2.5.3 Urban management dimension: conservation planning
    and management approaches ········ 048
  2.5.4 Social spatial dimension: gentrification,
    social spatial reconstruction and replacement ········ 054
  2.5.5 Sustainable development as an integrated approach ········ 056

Chapter 3 Theoretical Underpinning ········ 059
 3.1 Urban social sustainability ········ 059
  3.1.1 Social justice ········ 059
  3.1.2 Social cohesion ········ 062
 3.2 Social capital ········ 063
  3.2.1 Defining social capital ········ 063
  3.2.2 The nature of social capital ········ 065
  3.2.3 Forms of social capital: bonding, bridging
    and linking social capital ········ 067
  3.2.4 Social network structure ········ 068
  3.2.5 The adaptation of social capital theory ········ 070
  3.2.6 The potentials to adapt social capital approach
    in urban conservation decision making ········ 072
 3.3 Social relationship pattern in traditional society of China ········ 073
  3.3.1 Chinese social relationship structure CHA-XU-GE-JU (差序格局)···073

   3.3.2 Core values of Chinese society ——HE (和 /Cohesion)·················075

Chapter 4 Research Methodology·················076
 4.1 Case study approach·················076
 4.2 Research methods·················078
  4.2.1 Qualitative methods·················078
  4.2.2 Quantitative methods·················081
 4.3 Social capital survey·················082
  4.3.1 Survey design·················083
  4.3.2 Data collection, data entry and data editing·················087
  4.3.3 Examining research hypothesis 1 and hypothesis 2·················088
 4.4 Social network analysis under the framework
   of path dependency·················089
  4.4.1 Examining research hypothesis 3·················089
  4.4.2 Framework of path dependency·················090
  4.4.3 Social network analysis·················091
 4.5 Research framework·················094
  4.5.1 Summary of the research methodology·················094
  4.5.2 Main steps of the research·················094
  4.5.3 Research framework·················095

Chapter 5 Case Study of Hui Fang in Xi'an·················097
 5.1 Introduction of Ethnic Hui·················097
  5.1.1 The origins of Ethnic Hui in China·················097
  5.1.2 The origin of Ethnic Hui in Xi'an·················098
 5.2 The evolution of Hui Fang before1980s·················099
  5.2.1 Fan-fang in Tang Dynasty·················099
  5.2.2 Si-Fang in Yuan Dynasty·················101
  5.2.3 Hui-fang in Ming-Qing Dynasty·················102
  5.2.4 The structural change of Si-Fang in the period between 1912
    and 1949·················104
  5.2.5 Hui-Fang's Twelve-Mosque structure in the period between 1950s
    and 1980s·················105
 5.3 Urban fabric of Hui Fang community in Xi'an·················107
 5.4 The street life and social relationship in Hui Fang·················108
  5.4.1 Streets analysis·················109
  5.4.2 Street observation of Xiyangshi Street·················127

    5.4.3 Survey ········ 136
 5.5 A general introduction of urban redevelopment
    in Hui Fang after 1980s ········ 144
 5.6 The conservation process in Hui Fang ········ 147
    5.6.1 The initial stage (1980—2000) ········ 148
    5.6.2 The transitional stage (2000—2008) ········ 150
    5.6.3 The target stage (2008—2020) ········ 152
 5.7 A survey on local residents' perception towards Hui Fang's
    conservation ········ 153
    5.7.1 Residents' attitudes towards urban redevelopment ········ 153
    5.7.2 Changes of life in recent 10 years ········ 153
    5.7.3 Residents' attitude towards the Beiyuanmen Historic District Sino
       Cooperation Preservation Project (1997—2002) ········ 156
    5.7.4 Collective actions towards the Sajinqiao Redevelopment
       Project (2005—2007) ········ 160

Chapter 6 Case Study: Tianzi Fang in Shanghai ········ 163
 6.1 A brief introduction of current historic district conservation
    planning in Shanghai ········ 163
    6.1.1 Detailed regulatory planning for twelve areas with historical
       and cultural features ········ 163
    6.1.2 Urban conservation approaches and practices on historic quarters ········ 164
 6.2 Conservation process of Li-Nong neighborhoods in Shanghai ········ 166
    6.2.1 Conservation process ········ 167
    6.2.2 Current difficulties to conserve Li-Nong houses ········ 169
    6.2.3 The values of Shanghai Li-Nong houses ········ 170
 6.3 Urban conservation in Tianzi Fang ········ 171
    6.3.1 The evolution of Tianzi Fang (pre-conservation) ········ 171
    6.3.2 The development path of Tianzi Fang's conservation:
       from place making to place management ········ 174
    6.3.3 A survey on local residents' perception towards
       Tianzi Fang's conservation ········ 187

Chapter 7 Data Analysis ········ 199
 7.1 Community social capital based social cohesion analysis ········ 199
    7.1.1 Correlation analysis of variables within social cohesion indicator ········ 199
    7.1.2 Data extraction——factor analysis ········ 205

		7.1.3	Testing hypothesis 1 and hypothesis 2	207
	7.2	Social network analysis in the decision making process		212
		7.2.1	Hui Fang, Xi'an	212
		7.2.2	Tianzi Fang, Shanghai	216
		7.2.3	Testing hypothesis 3	223

Chapter 8 Conclusion ·········· 224

- 8.1 Research overview ·········· 224
- 8.2 Discussions of main findings ·········· 224
  - 8.2.1 Neighborhood social cohesion constructs in Chinese cities ·········· 224
  - 8.2.2 Neighborhood social cohesion and residents' attitudes of community conservation ·········· 225
  - 8.2.3 The dynamic compositions and interactions initiated by three forms of social capital in the decision-making network ·········· 225
- 8.3 A social capital integrated approach to urban conservation ·········· 226
  - 8.3.1 The place of social capital in the process of urban conservation of historic quarters in order to achieve social sustainability ·········· 226
  - 8.3.2 A social capital integrated approach to urban conservation ·········· 227
- 8.4 Contributions and implications ·········· 228
  - 8.4.1 Theoretic contributions ·········· 228
  - 8.4.2 Methodological contributions ·········· 228
  - 8.4.3 Practical implications ·········· 228
- 8.5 Limitations of the study ·········· 229
- 8.6 Conclusion ·········· 230

References ·········· 231
Appendix Principle Component Analysis (PCA) ·········· 262

List of Figures

Figure 4-1	Conservation decision-making process under path dependency framework	091
Figure 4-2	Research framework	096
Figure 5-1	Fan-Fang in Tang Dynasty	101
Figure 5-2	Twelve-Mosques pattern in Hui Fang	106
Figure 5-3	Prayer service in the Great Mosque	106
Figure 5-4	After Friday prayer service in Dapiyuan Mosque	107
Figure 5-5	Celebrating the Feast of the Sacrifice-Corban Festival in the courtyard of the Great Mosque	107
Figure 5-6	Urban fabric in Hui Fang	108
Figure 5-7	Typical courtyard house in Hui Fang	108
Figure 5-8	Selected streets and mosques for observation	109
Figure 5-9	Sajinqiao Street is now facing decay	111
Figure 5-10	Mix-used functional layout along Xiyangshi Street	112
Figure 5-11	Public facilities along Xiyangshi Street	113
Figure 5-12	Functional layout along Daxuexi Alley	115
Figure 5-13	Public facilities along Daxuexi Street	116
Figure 5-14	Functional layout along Guangming Alley and Beiguangji Street (northern part)	118
Figure 5-15	Functional layout along Beiguangji Street (southern part)	119
Figure 5-16	Public facilities along Guangming Alley and Beiguangji Street (northern part)	120
Figure 5-17	Public service along Guangming Alley and Beiguangji Street (southern part)	121
Figure 5-18	Functional layout along Dapiyuan Street	123
Figure 5-19	Functional layout along Xiaopiyuan Street——Residential Street	124
Figure 5-20	Public service along Dapiyuan Street	125
Figure 5-21	Public service along Xiaopiyuan Street	126
Figure 5-22	Pedestrian use in the early morning	129
Figure 5-23	Traffic congestion	129
Figure 5-24	Preparing food	131
Figure 5-25	Baby-sitting	131

Figure 5-26	Chatting	131
Figure 5-27	Giving in charity	131
Figure 5-28	Street vendors	131
Figure 5-29	Shopping	133
Figure 5-30	Dining	133
Figure 5-31	Cricket fight gambling	135
Figure 5-32	Chatting	135
Figure 5-33	Preparing food for gathering	135
Figure 5-34	Street watching	135
Figure 5-35	Funeral	136
Figure 5-36	Wedding	136
Figure 5-37	Religious festival	136
Figure 5-38	The subdivision of courtyard house within three generations	145
Figure 5-39	Self construction by local residents in Huajue Alley	145
Figure 5-40	The transformation of street elevations of Xiyangshi Street from 2000 to 2012	146
Figure 5-41	The self-construction along Daxuexi Alley	147
Figure 5-42	The self-construction between North Guanji Street and Chenghuang Temple	147
Figure 5-43	Planning of Beiyuan men Street, 1983	149
Figure 5-44	Renovated courtyard house No.125 in Huajue Alley	150
Figure 5-45	Scope of BHD in Xi'an Master Plan (1980—2000)	150
Figure 5-46	Scope of BHD in Proposed Xi'an Master Plan (2004—2020)	151
Figure 5-47	The Eleven mosques' joint seals to establish the Mosque Union and one page of local residents' joint signature in the petition to the central government against demolition	152
Figure 5-48	Physical deterioration and functional decline of Sajinqiao Street	154
Figure 5-49	Medical and educational resources in HF in 1997	155
Figure 5-50	Medical and educational resources in HF by 2013	155
Figure 5-51	Abandoned Sajinqiao Preliminary School	156
Figure 5-52	Abandoned Huajuexiang Preliminary School	156
Figure 5-53	Abandoned Tuanjie Preliminary School	156
Figure 5-54	Abandoned Huis Hospital	156
Figure 6-1	Twelve historical and cultural areas	164
Figure 6-2	French concession area	172
Figure 6-3	Taikang Road historic area	172
Figure 6-4	Taikang Road area	172
Figure 6-5	Mix-used neighborhood, Li-Nong and neighborhood factory	173
Figure 6-6	*Nongtang* styled fabric and diverse combination.	173

Figure 6-7	Taikang Art Street	175
Figure 6-8	Indoor market "Yilufa" which is renovated from Li-Nong neighborhood factory	175
Figure 6-9	Artist factory——an adaptive reuse of Li-Nong factories	176
Figure 6-10	The demolished Li-Nong neighborhood Xin-Xin Li	177
Figure 6-11	Shared kitchen	178
Figure 6-12	Dilapidated facilities	178
Figure 6-13	Cramped living space	178
Figure 6-14	Xin-Xin Li Redevelopment Plan (2003) which includes northern and southern parts	179
Figure 6-15	Xin-Xin Li Redevelopment Project and Taikang Road Neighborhood in the north	180
Figure 6-16	Detail Plan of Xin-Xin Li area 2003 and the conserved factories area (red marked buildings) in 2004	180
Figure 6-17	Document for conserving the factories in 1st step, 2004	181
Figure 6-18	functional reuse of residential units in Tianzi Fang	182
Figure 6-19	Functional reuse of Er Jing Alley	183
Figure 6-20	House owners signed jointly to protect the Intellectual Property	183
Figure 6-21	Tenants signed jointly an agreement to protect the Intellectual Property	184
Figure 6-22	The agreement to protect to protect the Intellectual Property	184
Figure 6-23	Physical improvement in Tianzi Fang	184
Figure 6-24	Land use of Tianzi Fang Neighborhood adjusted from residential to commercial-residential land and removed from redevelopment category	185
Figure 6-25	Spatial transformation of Tianzi Fang from 2004 to 2009	192
Figure 6-26	Functional layout of Tianzi Fang in 2013	192
Figure 6-27	Creative use of historic buildings	193
Figure 6-28	Commercialized Tianzi Fang	193
Figure 6-29	Space of unsatisfactory residents	193
Figure 6-30	Living narratives	194
Figure 6-31	Tianzi Fang in 2007	195
Figure 7-1	Network analysis of XMHDPP 1997—2002	215
Figure 7-2	Network analysis of SRP 2005—2007	215
Figure 7-3	Initial stage of the conservation process in Tianzi Fang	219
Figure 7-4	Transitional stage of the conservation process in Tianzi Fang	220
Figure 7-5	Target stage of the conservation process in Tianzi Fang	220

List of Tables

Table 4-1	Main interviewees during fieldwork of Hui Fang, Xi'an	079
Table 4-2	Main interviewees during fieldwork of Tianzi Fang, Shanghai	081
Table 4-3	Social cohesion indicator of Benevolence	083
Table 4-4	Social cohesion indicator of Righteousness	084
Table 4-5	Social cohesion indicator of Propriety	084
Table 4-6	Social cohesion indicator of Wisdom	085
Table 4-7	Social cohesion indicator of Faithfulness	085
Table 4-8	Structural social capital as cohesion indicator	086
Table 4-9	Local residents' general attitudes of conservation as a special indicator	086
Table 5-1	Demographic profile of respondents	137
Table 5-2	Socio-economic profile	138
Table 5-3	Survey about Hui Fang——mosque and community committee	140
Table 5-4	Survey about tradition	141
Table 5-5	Residents' perception towards Hui Fang's change and business activities	142
Table 5-6	Culture and sense of place	143
Table 5-7	Culture relics and buildings in Hui Fang	148
Table 5-8	Life change in recent 5 years	154
Table 5-9	Interview of residents' attitudes towards the Beiyuanmen Historic District Sino Cooperation Preservation Project (1997—2002)	157
Table 5-10	Interview of collective actions towards the Sajinqiao Redevelopment Project (2005—2007)	160
Table 6-1	The comparison of three cases	164
Table 6-2	The value of Li-Nong houses changes over time	170
Table 6-3	Demographic profile	188
Table 6-4	Socio-economic profile	188
Table 6-5	Change of dwelling in recent 5-10 years	189
Table 6-6	Resident's attitudes towards future redevelopment	196
Table 6-7	Residents' perception on the quality of life	197
Table 7-1	Correlations analysis of variables of social cohesion indicators in Hui Fang, Xi'an	199
Table 7-2	Correlations analysis of variables of social cohesion indicators	

	in Tianzi Fang, Shanghai	202
Table 7-3	Correlations analysis of social cohesion constructs and resident's attitudes towards conservation in Hui Fang, Xi'an	207
Table 7-4	Correlations analysis of social cohesion constructs and resident's attitudes towards conservation in Tianzi Fang, Shanghai	208
Table 7-5	Nodes producing social capital to Hui Fang during Xi'an Muslim Historic District Protection Project (1997—2002)	213
Table 7-6	Nodes producing social capital to Hui Fang during Sajinqiao Redevelopment Project (2005—2007)	214
Table 7-7	Density overall of the conservation process in Hui Fang, Xi'an	216
Table 7-8	Degree centrality of the conservation process in Hui Fang, Xi'an	216
Table 7-9	Nodes producing social capital to Tianzi Fang in initial stage	217
Table 7-10	Nodes producing social capital to Tianzi Fang in transitional stage	218
Table 7-11	Nodes producing social capital to Tianzi Fang in target stage	219
Table 7-12	Density overall of the conservation process in Tianzi Fang	221
Table 7-13	Degree centrality of the conservation process in Tianzi Fang	221

List of Charts

Chart 5-1　Transportation activities of Xiyangshi Street, 20-9-2012 (Thursday) ············ 128

Chart 5-2　Transportation activities of Xiyangshi Street, 22-9-2012 (Saturday) ············ 128

Chart 5-3　Transportation activities of Xiyangshi Street, 24-9-2012 (Monday) ·············· 128

Chart 5-4　Transportation activities of Xiyangshi Street, 9-10-2012 (Tuesday) ············ 128

Chart 5-5　Activities of shop owners in Xiyangshi Street, 20-9-2012 (Thursday) ············ 130

Chart 5-6　Activities of shop owners in Xiyangshi Street, 22-9-2012 (Saturday) ············ 130

Chart 5-7　Activities of shop owners in Xiyangshi Street, 24-9-2012(Monday) ············ 130

Chart 5-8　Activities of shop owners in Xiyangshi Street, 9-10-2012 (Tuesday) ············ 131

Chart 5-9　Activities of tourists in Xiyangshi Street, 20-9-2012 (Thursday) ················ 132

Chart 5-10　Activities of tourists in Xiyangshi Street,22-9-2012(Saturday)················ 132

Chart 5-11　Activities of tourists in Xiyangshi Street, 24-9-2012 (Monday) ················ 132

Chart 5-12　Activities of tourists in Xiyangshi Street, 9-10-2012 (Tuesday) ················ 133

Chart 5-13　Activities of residents in Xiyangshi Street, 9-20-2012 (Thursday) ·············· 134

Chart 5-14　Activities of residents in Xiyangshi Street, 9-22-2012 (Saturday) ·············· 134

Chart 5-15　Activities of residents in Xiyangshi Street, 9-24-2012 (Monday) ················ 134

Chart 5-16　Activities of residents in Xiyangshi Street, 10-09-2012 (Tuesday)·············· 135

Chart 6-1　Income change in the last 5-10 years ··· 190

Chart 6-2　Current dwelling condition ··· 197

Chapter 1 Introduction

Developed from architectural preservation, urban conservation has now become increasingly important for cities. Especially in cities under rapid development, urban heritage conservation has now been frequently used as an urban tool to revive the decayed urban areas under various contexts such as economic restructuring, industrial relocation, rapid urbanization, globalization, and so on and so forth. Under these transitional social and economic contexts, recent decades have witnessed the evolving values of heritage,the shifting paradigms of urban conservation doctrines, and the emerging appeals for sustainable development, especially the dimension of social sustainability. With the intention to examine how social capital might contribute to urban conservation by filling in the gap of urban conservation's role in the social sustainability with regards to two key indicators of social values——social cohesion and social justice, this study will discuss the roles of social capital in urban conservationby using a mixed methodology, which includes case study, social capital survey and social network analysis under a decision making framework of path dependency. This chapter presents a brief introduction of urban conservation, which includes the following sections: significance of urban conservation,the research background and context, research statements, potential contribution of this research, the research scope, and an outline of the subsequent thesis chapters.

1.1 Significance of urban conservation in urban development

Urban conservation has now been used as an urban tool not only to help a city retain its historic and cultural qualities——in terms of symbolism imagery, and aesthetic value——but also to assure a city the continuity of (1) urban fabric, patterns and structure; (2) traditional technology and material and the connections of memories; and (3) feelings of shared identity among the people, all of which in turn will improve the quality of life in a city(Orbasli, 2000; Pendlebury, 2008; UNESCO, 2011). Nowadays urban conservation has become progressively more important for cities, particularly for cities experiencing rapid development and change.

The "bulldozer" mode of mass demolition and mass construction has been recurrently implemented in many historic quarters all over the world, which led to myriad impacts on cities in the long run. The irreversible destruction of human and spiritual values(Huntington, 1971), accompanied with the growing environmental awareness under dramatic changes has led to the feelings of "uncertainty"and "a lost world" and raised a number of resistances of demolition

(Pendlebury, 2008). The field of material fabric based conservation——that is architectural preservation——has been criticized to have isolated historic buildings in a museum-like quality without considering the needs of locals,thus devaluing the significance of location and economic potentials. In addition, together with the process of rapid modernization and global homogenization[①] in many cities, the city image and urban experience are becoming increasingly similar from city to city which might lead to a "less attractive"urban experience (Harvey, 1989). Furthermore, the voices appealing for cultural continuity, diversity,and identity have never been louder. As urban conservation is increasingly influenced by other urban categories, scholars argue that conservation should be an integral part of urban planning(Ashworth & Tunbridge, 1990; N. Cohen, 1999; Hobson, 2004; Tiesdell, Oc & Heath, 1996).As such, urban conservation has been increasingly involved into the domain of modern urban planning since late 1960s (Pendlebury, 2008).

With the economic restructuring and industrial relocation on a worldwide scale, many historic quarters have lost their original functions and decayed over time (Larkham, 1996; Lichfield, 1988; Tiesdell, et al., 1996). The economic restructuring process has transformed historic quarters from functional "self-sufficient" ones to those largely dependent on the economic dynamism of the city, which urges the exigency of economic revitalization in historic quarters (Tiesdell, et al., 1996). Globalization and an increasingly internationalized economy have spurred transnational cooperation and competition between cities (Castells & Hall, 1994). In many cities, historic areas are searching for new functions and activities to revive the dynamic economy and are *repositioning themselves into the global economy* " under a widely adopted marketing process of a heritage-led "place marketing" (Haughton & Hunter, 1994, p. 39). In this sense, historic quarters become part of the economic dynamics of the city (Tiesdell, et al., 1996; Urry, 1995a), and heritage has been largely commodified (Pendlebury, 2008). Accordingly, urban conservation has evolved from a regulation-based technical planning tool of preserving cultural and historical values to a policy led urban tool for boosting economic growth and urban development(UNESCO, 2011; Urry, 1995a). For instance, tourism(Ashworth & Tunbridge, 1990; Orbasli, 2000), housing(Pendlebury, 2008; Tiesdell, et al., 1996), and the cultural industries(Gibson & Kong, 2005; Landry, 2008; Murray, 2001) are popular initiatives that align with the latter motive. Given the ever-increasing relevance to urban development, urban conservation has become one of the core issues in urban planning to manage urban development.

Due to the limitations of modern planning, in dealing with urban heritage which are manifested

---

① Huntington (1971) identifies "revolutionary", "complex", "systemic", "global", "lengthy", "phased", "homogenizing", "irreversible" and "progressive" to be nine key characteristics of the modernization process, especially among which, "revolutionary", "global", "homogenizing" and "irreversible" have caused dramatic changes to cities.

as both tangible and intangible heritage, orthodox urban conservation prefers to employing top-down approaches and implementing "hard" measures such as technical regulations without considering the impacts on local communities and individual's everyday life, thereby often leads to social conflicts.

As urban development, understandably, is a complicated process, urban conservation has been consistently intertwined with shifting urban contexts and many other driving forces, hence being an ever-evolving doctrine with continuous methodological adaptations to reconcile the emerging conflicts and the rising agendas. Ever since late 1970s and 1980s, the cultural pluralism and diversity based on postmodern perspective which criticized the economic and social consequences of urban conservation practices such as gentrification led social replacement have given rise to new agendas for urban conservation to reconcile such as loss of community identity and collective memories(L. Smith, 2006; Urry, 1995b), continuity and frequency(Pendlebury, 2008; UNESCO, 2011), everyday life heritage (Wells, 2007), affordability and social diversity (Atkinson, 2000), etc.

For instance, conservation of urban historic quarters and neighborhoods has become progressively more significant to sustain the urban social ecology. The condition of a historical quarter[①], decay or growth, can be seen as a result of many influential factors, such as political forces, economic activities, planning intervention, industrial relocation, functional change, social spatial change, accessibility obsolescence, and so on and so forth. Since late 1960s, community resistances of the profit-driven redevelopment of historic areas have been happening recurrently all over the world when higher socioeconomic groups, who are served by high-end development, replace existing lower-income social groups and their traditional mixed-use neighborhoods, particularly in contemporary Chinese cities, where urban development is extremely rapid, and mainstream urban conservation practices favor the "demolition" approach through a top-down process.

On one hand, the spontaneous process of socializing is interrupted by one-time mass demolition and relocation of original residents to the urban fringe areas where employment opportunities and civil infrastructures are inadequate, which reduce the quality of life and the affordability of cities, and increase the burdens on cities in the long run. On the other hand, social replacements diminish the social values attached to the physical fabric of neighborhoods and harm the openness and inclusion in a society.

---

① Historical quarters or settlements are generally small-sized, mixed-use urban areas with a manifestation of the traditional urban fabric.

Built on the above discussions, urban conservation could be understood asa significant process of urban development in terms of cultural significance, economic boost and social construct. Nowadays, by putting urban development into a big image, sustainable development which aims to solve the problem of contemporary development without compromising the future of successive generations has attracted worldwide attention from all levels: government, academics, institutions and individuals. In response to the changing social economic situations worldwide, there urged a call for urban conservation to adopt a sustainable perspective, as stated by UNESCO(2011, p.3) *"As the future of humanity hinges on the effective planning andmanagement of resources, conservation has become a strategy to achieve a balance between urban growth and quality of life on a sustainable basis."*

As such, the understandings of urban value embedded in the urban heritage should also evolve to be identified and redefined collectively according to the angle of sustainable development (ICOMOS, 2010; Nyseth & Sognnæs, 2013; UNESCO, 2011). Nevertheless urban conservation, which should accordingly addresses "legacy of past", contemporary as well as future development, has not yet fully evolved, especially in the dimension of urban social sustainability, such as social inclusion and social justice(Pendlebury, 2008; UNESCO, 2011).

1.2 Research background and context

1.2.1 Research background——urban conservation and sustainable development

Developed since the 19th century, along with the ever-evolving understandings of heritage under the constantly changing political,economic, social and cultural contexts, urban conservation has now been widely recognized as a collection of heritage value centered place management processes to manage change (Avrami, Mason & Torre, 2000; Madgin, 2009; Pendlebury, 2008; Tiesdell, et al., 1996; UNESCO, 2011; Wells, 2007). While the core role of the discipline of urban conservation remains to retain the values imbued in urban heritage, the scope of the field has expanded from architectural preservation to the place management processes, and the understanding of "change" has transcended the physical dimension towards the multiple dimensions.

Current study of urban conservation has developed a variety of approaches to evaluate and identify heritage values and to manage urban conservation process, which can be categorized in historic and aesthetic dimension, economic dimension, conservation planning and management dimension, social spatial dimension, and so on and so forth.

In the cultural and historical dimensions for historical sites efforts are taken to explore the cultural, historic and aesthetic value of conservation substances: An area-based approach was adopted to enlarge the scale of conservation objects from single historic monuments to groups of buildings, markets and towns, etc.(Hobson, 2004; Larkham, 1996; Rodwell, 2007; Tiesdell, et al., 1996).Adapted by many researchers, the townscape has become a powerful instrument to manage and instruct urban conservation and renewal (N. Cohen, 1999; M. Conzen, 1973; Cullen, 1961; Geddes, LeGates, & Stout, 1949; Slater, 1984; J. Whitehand, 1990). With the growing research interest of conserving the identity of urban areas, urban morphology has been adopted to understand the hidden structure and relationship of various elements within a historic area, and in the identification of layers of historical development(M. R. G. Conzen, 1960; Gu 2001; J. Whitehand, Gu, Whitehand & Zhang, 2011; J. W. R. Whitehand, 1992; Widodo, 2004). Approaches in this dimension are mainly focusing on physical aspects hence ignoring the intangible value of heritage.

Approaches in economic dimension explore the following topics such as revitalization of historic quarters against all kinds of obsolescence (Larkham, 1996; Lewis & Weber, 1965; Lichfield, 1988; Mason, 2005; Sim, 1982; Tiesdell, et al., 1996), functional reuse historic areas and commodification of heritage (Hobson, 2004; Pendlebury, 2008; Pickard, 2001; Tiesdell, et al., 1996), reposition of historic centers under economic restructuring(Castells, 1985; Harvey, 2003; Haughton & Hunter, 1994; Tiesdell, et al., 1996; Urry, 1995a), the economic analysis on demolition vs. functional reusesof historic remains(Hobson, 2004; Madgin, 2009; Pendlebury, 2008; Pickard, 2001; Tiesdell, et al., 1996; S. Zhang & Chen, 2010), and economic value evaluation (Mason, 2005). In general, these approaches lack the concerns of cultural, social and public aspects, thereafter have led conservation into an unbalanced situation: economic revitalization and consumption of heritage (Ashworth & Tunbridge, 2000; Pendlebury, 2008; Urry, 1995a)become the prime concerns in decision-making and policy-making.

In the dimension of conservation planning and management, conservation planning control guidelines and regulation provide a technical, regulatory and relatively top-down approach for conservation implementation. It is easy to adopt, while it lacks the historic and contextual view provided by urban morphological analysis and townscape study (Barrett, 1993; Hobson, 2004; Larkham, 1996; Mageean, 1999; Pickard, 2001).The following approaches can be characterized as "soft" approaches, which include conservation planning integral policy(Madgin, 2009; Pendlebury, 2008; Strange, 1997; UNESCO, 2011), political support and public intervention(Madgin, 2009; Tiesdell, et al., 1996), place management(Healey, 2006; Hobson, 2004; Kearns & Philo, 1993; Murray, 2001; Tiesdell, et al., 1996; Worthington, Warren & Taylor, 1998), public participation(Atlee, 2008; Creighton, 2005; Ping, 2007; Sharif Shams, 2006; Y. Zhang, 2008),and conservation-initiated community development(Bhatta, 2009; Elsorady, 2011; Pendlebury, 2008; Salazar, 2006; Snyder, 2008; Wells, 2010). Particularly by integrating community needs and resources into conservation planning, as a bottom-up approachthe community development approachaims to provide multiple development choices with fewer impacts on community identity and daily life (Salazar, 2006). These soft approaches could, to a certain extent, promote participation and understanding of sense of place, compromise the top-down approach, and balance the conflicts among different interest groups(Creighton, 2005; Larkham, 1996; Strange, 1997; Tiesdell, et al., 1996). As there commonly lack of a comprehensive communication and associational framework, hence reducing the efficiency of conservation in terms of negotiating decision making, plan implementation, and so on.

The social spatial dimension of urban conservation employs the approaches of gentrification and social replacement(Appleyard, 1979; Atkinson, 2000; Gottdiener & Budd, 2005; Lees, Slater & Wyly, 2010; N. Smith, 1979; Tiesdell, et al., 1996), and social spatial reconstruction (Gottdiener & Budd, 2005; Gottdiener & Hutchison, 1994; Gratz, 1995). These approaches provide perspective in understanding the economic, social, and political context of urban development, and examine the limitations of urban conservation as atool to manage change under the dominant concern of economic gains which undervalues the social diversity and cultural identity of local communities(Urry, 1995b). Nevertheless, this approach strongly emphasizes the function of private capital and the free market and overlooks the function of urban planning policy and management, thus failing to provide the further solutions for the identified social issues.

The above approaches have explored the field of urban conservation in both doctrine and methodological approaches. As a heritage value centered place management process to manage change, the development of urban conservation has been closely interrelated with the evolution of the understanding of heritage values and the shifting contexts. In this sense, the above approaches to study urban conservation could also reflect the way how we understand heritage values.

In the past two decades, under the changing political, social and economic contexts the postmodern critiques based on cultural relativism urge the identification of heritage values to address cultural plurality, diversity, and individual's perspective on heritage values.

Nevertheless, current conservation doctrine is criticized to "codify, fixate and prevent the evolution of alternate meanings"(Wells, 2010, p. 481) and lack of both doctorial guidance and methodological instrument in value identification (Avrami, et al., 2000; Pendlebury, 2008; Wells, 2007).In addition, compared to other dimensions, literature review shows there lacks sufficient approaches to address the social dimension of heritage values.

Furthermore during recent years, sustainable develop has become a worldwide consensus which put urban conservation into a big image. The agenda of sustainable development has provided a new avenue for heritage value to evolve and for urban conservation to adapt, which means that urban conservation ought to be more than merely physical maintenance of a historic site but also the heritage value centered management of place from a process perspective towards sustainability. In other words, a sustainable development perspective to urban conservation should ask: How can conservation planning help to manage the process of change and development in historical places, which will lead to a sustainable transformation in the long run?

In order to better understand urban conservation and to cope with the shifting views of heritage value, sustainable development could be understoodas a new approach to conservation planning which highlights the long-term dynamic integration and cooperation of different approaches based on different understandings of values from different social groups and individuals.Conservation as the process of managing change could be examined in that whether the development path of a historic place under diverse influential factors and driving forces will be ultimately towards sustainability, or in other words, whether different value identifications from different social groups and individuals could negotiate the decision making of each development stage of a historic area and present a trend toward sustainable development.

Nevertheless current studies of sustainable urban conservation in historic quarters focus predominantly on aspects of the economy(Al-hagla, 2010; Nasser, 2003; Tweed & Sutherland, 2007) and environment(Brimblecombe & Saiz-Jimenez, 2004; Savage, Huang & Chang, 2004). In aspects of socially sustainable urban conservation, a few studies have been conducted on identifying collective memories as indicator to drive urban conservation towards sustainable development (Ardakani & Oloonabadi, 2011), on addressing the role of collaborative forms of governance to the inclusive preservation of historic towns (Nyseth & Sognnæs, 2013) , and on understanding the heritage value identification as a cultural practice led social process which should be inclusive and active with diverse groups of people participate in (Pendlebury, 2008, p. 37).

It has been verified in many cases that without a solid social network and social justice, urban conservation could not achieve sustainability, and the economic and environmental dimensions would decline progressively (Mo & Lu, 2000; RUAN & YUAN, 2008; Shao, Zhang & Dun, 2004). However, according to current studies on urban conservation, sustainability concerns, particularly the social dimensions of sustainability, are largely missing in the body of knowledge surrounding urban conservation in historic quarters, and there is a certain lack of feasible approaches to promote social sustainability,which raise the critical question about its role and significance in the conservation planning process.

In an effort to strengthen the social dimension in sustainable development, Putnam (2000) recommends that public policy makers should pay more attention to the formation of social capital. Cuthill (2010) also argues that "social capital provides a theoretical starting point for social sustainability." Accordingly, understanding social capital theory and the adaptationsare essential in strengthening social sustainability in urban conservation.

Social capital is thought to be the "missing link" in urban development and "essential glue" for the society (Halpern, 1999). Social capital is understood as resources and assets of relations (Bourdieu, 1986; Coleman, 1994), network structures (Coleman, 1994; Lin, 2002),and trust and norms (Fukuyama & Institute, 2000; R. D. Putnam, Leonardi & Nanetti, 1993). Social capital as a public good will benefit individuals, group members, and the community, and lead to collective action and cooperation (Coleman, 1994; Pena, Lindo-Fuentes & Unit, 1998). Rather than a one-time feat, social capital is often produced in a process of repetition where the norms of reciprocity and trust emerge in light of obligation and expectation (R. D. Putnam, et al., 1993). As individual trust is transformedinto social trust, which resolves the dilemma of collective action under the logic of rational choice, transaction costs are reduced, and the efficiency of governance is also increased(Pena, et al., 1998; R. D. Putnam, et al., 1993).

In addition, social capital would have the potential to be converted into economic,cultural (Bourdieu, 1986), and human capital (Coleman, 1994). Putnam, et al. (1993) identified two ideal forms of the social network structure: horizontal and vertical. The three forms of social capital are categorized as bonding, bridging, and linking social capital (R. Putnam, 2000). Bonding social capital consists of "the strong ties connecting family members, neighbors, close friends, and business associates" (Bank, 2000). Bridging social capital can "bring together people across diverse social divisions" (Scott, 1991) and "generate broader identities and reciprocity" (R. Putnam, 2000). Linking social capital is formed by "vertical connections that connect individuals and groups with institutions" (Babaei, Ahmad, & Gill, 2012).

With regards to its adaptations,the research scope of social capital has extended from sociology and politics to many other fields, such as alleviating poverty among vulnerable social groups and poor communities(Bank, 2000; Woolcock, 1998), economic enhancement among informal economic groups(Coleman, 1994; Fukuyama & Institute, 2000; Woolcock, 1998), solving economic development dilemmas (Woolcock, 1998), community governance (R. D. Putnam, et al.,

1993), institutions and organizations and policy making (D. Cohen & Prusak, 2001; Ozmete, 2011; Woolcock, 2001), promoting social cohesion(World Bank, 2000) and so on.

In policy making, the dynamic combination and interaction between the three social capital forms have been highlighted (Fukuyama & Institute, 2000; Woolcock, 2001). Social capital highlights the process rather than the outcome (Woolcock, 2001), and its adaptation of policy helps achieve collective action and reciprocity, promoting dynamic relationships between networks. There exist potentials in adapting a social capital approach in urban conservation under a decision-making framework of path dependence. An observation of the different roles and combinations of social capital might reveal how decision-making was influenced and retargeted in the urban conservation process.

To return to the field of urban conservation, there lacks a decision-making framework of identifying heritage value in that "the reasons why and what, we seek to conserve and the implications that lead from this as to how we conserve", and of negotiating "whose heritage is it"(Pendlebury, 2008, p. 214).In addition, as is pointed in the Getty report, "broadly, we lack any conceptual or theoretical overviews for modeling or mapping the interplay of economic, cultural, political, and other social contexts in which conservation is situated"(Avrami, et al., 2000, p. 10). Based on the above insights of social capital, this inspires the motive to examine the role of social capital in the research area of urban conservation and explore avenues through which social capital can be adapted within the current body of knowledge on urban conservation.

1.2.2 Research context——Chinese cities

In Chinese context, economic, historical, and architectural values of urban conservation have been well recognized, but its social value which can contribute to social harmony and cohesion is ignored, as can be revealed from the following challenges of urban conservation in China.

1) Challenges from rapid urbanization and economic development

After reform and opening-up policy, Chinese development of market economy has achieved remarkable results which have attracted worldwide attention. Nonetheless, it may also threaten urban conservation. According to Chiu and Moss (2007), market economy mechanism would focus on the maximizing the economic profit potential in short term but pay less attention to other kinds of value. There are a significant number of cases in China in which local governments and developers demolished historic areas and redeveloped them as gated communities and commercial districts (Shan, 2007).

In recent decades, the urbanization rate in China increases very rapidly.Chiuand Moss(2007) argued that reconstruction and redevelopment of historic areas are popular means of urbanization in China due to their economic value and the lack of land for new development. In the process of reconstruction, quite a number of traditional buildings were demolished in order to provide space for new projects (Ibid.).

On one hand, this phenomenon reflects that the economic value of historic areas is well recognized; but on the other hand their historic and architectural values are neglected on some level. Fortunately, central government has announced a series of laws and regulations to preserve historic areas and cities, which have been introduced in the previous section. Meanwhile, besides the increase of economy development, the situation of urban conservation has also become one of the evaluation factors to assess the achievement of local government and cities' competitiveness (Mou & Tan, 2008).

2) Challenges from modernization and globalization

With no doubts, the tendency of globalization and modernization has swept all over the world. In China, this idea of modernization can be traced back to the 18th century, which is to adapt western culture and products into people's daily lives (Wu, Katz, & Lin, 2010). For a long period, modernity could be found in political, economic, social, and environmental aspects as an important objective for development in China. However, according to Chen (2011, p. 411), "in the current Chinese context, it is a challenge to incorporate tradition within the modernization process and to achieve the consistent order". Similarly, globalization also forces Chinese traditional culture to interact with western culture (Ibid.). Consequently, historical regeneration often leads to gentrification and culture homogeny might become inevitable (Chiu & Moss, 2007).

In the past few decades, due to misunderstanding of the relationship between tradition and modernization, most of Chinese cities adopted modern architecture as dominated form to redevelop urban historic areas (Chiu & Moss, 2007). Due to limited understanding of the historical value of historic areas, "constructive destruction" and "destructive construction" could be found in the process of urban conservation, and as a result, local identities, historic memories, and city characteristics were lost (Ibid.).

According to Chen (2011, p. 411), tradition refers to "something inherited and transmitted over generations with clear continuity" and traditional urban forms reflect "residents' self-identity and are evolving according to their needs". Emphasized in central government's guidelines, traditional characteristics shall be sustained and persevered with appropriate urban development. Obviously, traditional urban form is the carrier of cities' historical values and should be well conserved. However, for local governments, getting a clear understanding of the relationships between modernity and tradition, globalization and local identity may need a gradual process.

3) Challenge from incomplete conservation mechanism in planning process

Although multi-dimension conservation system which consists of Cultural Relics Protection Units, historic areas and towns, and historic cites has been set up by central government, there are still some components are missing. Among them, the apparent lack of attention on the social value of urban conservation might be the most significant one.

Although research appeals to introduce public participation in the process of urban conservation in order to ensure local residents' rights and interests, there is no clear guidance regarding it given by governments (Guo, 2012; Wang, 2004; Xu, 2007). As introduced in previous sections, relocating all or part of local residents is a commonly adopted approach in the process of urban conservation and regeneration. Nonetheless, this approach may undermine the social network and cohesion in historic neighborhoods, which is built up through several generations.

1.3 Statement of the research topic

Upon re-addressing the ideology of urban conservation in that values including social values remain to be the core of heritage conservation, it is in the same time imperative to understand conservation as a process to manage change. However in determining "what to conserve" in the heritage value selection, current studies often disregard the important role that stable networks-splay in the development of cohesive social relationships and everyday activities within local communities which contribute to the vitality of the place and a sense of belonging. In terms of conservation approach, current planning and economic approaches favor technical planning approaches with a top-down strategy that advocates economic revitalization with an absence of social revitalization. The "soft" approach which utilizes both informal and formal networks to promote sustainable conservation is less exercised. The social-spatial approach lacks a deep understanding of the functions of urban policy and management in the process of intervention. A community development approach focusing on community issues does not adequately establish a network with both vertical and horizontal links to exterior parts. In terms of evaluating the outcomes, current research on urban conservation emphasizes the economic gains over the loss in other aspects, especially when the issues of social cohesion and spatial injustice are addressed. In addition, as urban conservation case studies differ from one another due to specific political, economic and cultural contexts, the study of sustainable urban conservation in the context of contemporary Chinese cities remains to be investigated at length.

1.3.1 Identification of the knowledge gaps

Sustainable development requires a holistic thinking and a new perspective. Sustainable urban conservation should be evaluated not only by economic gains and technical planning standards, but also by the mitigation of environmental and social impacts. Thus, through the process of urban conservation, historic quarters should be sustainably developed and economically, environmentally and socially transformed. Especially in terms of social sustainability, social cohesion and social justice are identified as key indicators(World Bank, 2000; Cuthill, 2010), that bury significant social values and should not be ignored or damaged by urban conservation.According to current studies on urban conservation, sustainability concerns, particularly the social dimensions of sustainability, are largely missing in the body of knowledge surrounding urban conservation in historic quarters.In addition, there is a certain lack of feasible approaches to promote social sustainability in the conservation process.

Social capital is thought to be the "missing link" in urban development and an "essential glue" for society (Halpern, 1999). The study on social capital theory and its adaptation in other research fields provides insights for socially sustainable urban conservation. The theory of social capital has been discussed widely and adapted in many areas such as public health(Kawachi & Berkman, 2000; Lomas, 1998; Lynch, Due, Muntaner, & Smith, 2000), economic growth(World Bank, 2000; Woolcock, 1998), and industrial cluster(Baker, 2000), income distribution(Glaeser, Laibson & Sacerdote, 2002), business and enterprise management(Baker, 2000; HÜppi & Seemann, 2001),etc.. Yet social capital as an approach to urban conservation is a marginally discussed topic. The impact and contribution of social capital to the sustainable conservation of historic quarters is yet to be clear. One would argue, given the knowledge gaps identified above, it is necessary to examine the place of social capital in the research area of urban conservation, and to find out how to adapt social capital into the current body of knowledge on urban conservation.

1.3.2 Research questions

Based on the identified knowledge gaps, which are the impact and contribution of social capital to the sustainable conservation of historic quarters and the adaptation mechanism of social capital into the current body of knowledge on urban conservation, this research proposes the research questions as the following:

What is the place of social capital in the process of urban conservation in historic quarters in order to achieve socially sustainable development in Chinese cities?

Given that social sustainability could be indicated through the examination of social cohesion as social value and social justice as an impetus of transformation, the research questions can be further developed into the followings sub-questions:

- Research question 1:

What are the general social capital indicators formed by stable social relationship, that are unique for constructing community collective social value——community social cohesion in traditional dwellings of China cities?

- Research question 2:

How can community social cohesion based social relationships influence residents' perceptions towards community conservation?

- Research question 3:

Based on the different decision-making process of the two selected case studies in China, what can be discerned from the patterns of composition and interaction of different forms of social capital, which could provide implicates for urban conservation policy making?

1.3.3 Research hypotheses

To answer the above research questions, the research hypothesizes that:

Hypothesis 1:

In traditional Chinese communities, there exist certain indicators of cognitive and structural community social capital, which are commonly important in maintaining neighborhood social cohesion, despite the diverse groups of residents.

Hypothesis 2:

Community social cohesion based on community social capital is a very influential factor on local residents' perceptions towards the conservation of traditional neighborhoods.

Hypothesis 3:

in the urban conservation decision-making network structure, there exist some patterns of dynamic composition and interactions of different forms of social capital, which are influential to break the lock-in effect in decision making, and to transform the conservation decision making towards a just decision-making process, thus creating new paths for urban conservation practices.

1.3.4 Research objectives

Given the dominant focus on the economic and environmental aspects of sustainable urban conservation in historic quarters, and lack of understanding on urban social sustainability, this book therefore aims:

- to evaluate whether the social capital can improve social cohesion and the maintenance of urban heritage using two selected cases.

- to investigate the dynamic composition and interactions of different social capital forms at different stages of the urban conservation process and to find out the implications for urban conservation policy.

- to propose a social capital integrated urban conservation approach for historic areas in urban China based on an inductive process, which could specifically enhance the dimension of social sustainability.

1.4 Potential contribution of this research

The findings of this study could extend the body of knowledge on urban conservation in terms of social sustainability through the adoption of social capital approach.On one hand,via social capital survey based on two cases, this study indicates that in the meso-level, through the long-term everyday interactions among social relationships within local community, social capital plays an important role in constructing the stable community social value——social cohesion, which has been identified as a key indicator of social sustainability, and yet has been largely neglected in current heritage value selection in urban conservation. On the other hand, through the social network analysis it is revealed that, as nodes of organizational relationship towards local community which could bring different forms of resources such as information, knowledge, materials, fund, power,and so on for potential developments, different forms of social capital——bonding, linking, and bridging social capital, eg.——interact and cooperate in the conservation decision-making process; and certain patterns which could contribute to transform or create the path towards inclusive decision-making pattern have been identified, which could contribute to social justice in a process perspective, hence promoting the dimension of social sustainability in urban conservation.

Based on the findings, this book proposes an alternative conservation approach, namely the social capital integrated urban conservation approach, which could contribute to urban conservation in terms of certain a practical implications.

1.5 Scope and scale of the study

1.5.1 Scope of the study

Due to the broad scope of heritage, which differs from the categories of tangible and intangible heritage, from natural heritage and cultural heritage, from the scale of individual building, groups of buildings, historic sites and historic areas, and also from functions such as residential quarter, industrial remains and cultural routes, etc., there might not exist a specific urban conservation approach which can be adapted universally. Firstly this study will focus on the urban heritage regarding to the scope of residential urban historic quarters, which are living urban heritage with local community involved, rather than merely physical historic relics. Secondly, aside from the scope of heritage, the strategy of urban conservation is closely related with urban management system, including level of urban governance, and urban policies such as social security and cultural development targets, etc.; needless to mention how influential the economic sector is. As this book aims to explore the role of social capital in terms of heritage value components and urban conservation decision-making process, this research will employ an inductive method on a case-by-case base, with each case possibly differing from the other in terms of political, economic, cultural, and social contexts. Accordingly the major focus would be on the mutual qualitative character of social capital among different cases of diverse urban contexts, rather than comparing the diverse urban contexts and analyzing thequantitative difference of social capital among respectively different cases.

To be more explicit, the mutual qualitative character of social capital refers to the following two aspects: firstly, as the cultural significance of heritage means "aesthetic, historic, scientific, social or spiritual value for past, present or future generations"(ICOMOS, 1999, p. 2), by which the social value matches the requirement of social sustainability, yet social value is usually discussed without attaching to specific contexts. Meanwhile in areas of social sustainability, social cohesion is highlighted as a significant aspect of the quality of societies(Berger-Schmitt, 2000) and particularly in a community level, social cohesion is pursued as most important value which could benefit community identity and bring people together(Uzzell, Pol, & Badenas, 2002). Accordingly social cohesion could be taken as a key indicator of social value, therefore on the meso-level which is community level, the qualitative social capital based on meaningful social relationship, for example trust and norms, etc. which facilitate to construct community social value——community social cohesion should be examined. In the meantime, it is also important to declare that within this meso-level, structural community social capital, such as network closure and network structure, informal organizations will not be a main focus of exploration.

Secondly, it is also addressed that social justice and equity are the two "ethical imperative" of social sustainability(Cuthill, 2010). Rather than evaluating social inequity in a perspective of outcome(Harvey, 1973), the social justice which embraces social inclusion as a key character could be examined in a process perspective(Cardoso & Breda-Vazquez, 2007; Fainstein, 2005; Young, 1990). Hence,with an effort to evaluate how social justice could be approached, this study goes further to examine in the meso-level organization based social capital to understand how different forms of social capital, namely bonding, linking and bridging social capital as organizational relationship resources have been inclusively involved in the entire decision-making process of urban conservation. Therefore social capital on the macro-level, which is municipal or national level, will not be discussed in this research.

Meanwhile, as social network analysis is employed in the analysis of decision-making process, it is also necessary to declare that this network analysis is based on the analysis of the interactions among organizational social capital on the meso-level decision-making platform, the social network within or beyond the meso-level of community decision-making platform will not be examined.

1.5.2 Scale of the study

The scale of the study is based on a level of urban historic area, and to be more specific, on Chinese cities through two heritage areas of two cities, namely Hui Fang in Xi'an and Tianzi Fang in Shanghai.Even though these two cases present different demographic profiles, and economic backgrounds, and development needs, the reason to focus on these two cases lies in that both heritage areas share the common bases, such as holding rich heritage characters and similar stable relationship among community, facing similar physical decay and development dilemma, which are both initiated by top-down approach, participated by different forms of social capital and transferred by different approaches, etc. Hence it is rationale to investigate the common roles of social capital to achieve social sustainability in both cases under the context of Chinese cities.

1.6 Outlines of the book

This book consists eight chapters. Chapter One introduces briefly the topic of urban conservation via the research backgrounds and contexts, presents the statements of research topic, and then closes with the declaration of research significance and scope. Both Chapter Two and Chapter Three represent the work of literature review. Chapter Two introduces the study of urban conservation in terms of definition, its development, and practices, and by concluding the current research studies and research areas of urban conservation, it highlights thatsustainable development especially in the social dimension requires new approach of urban conservation. Chapter Three firstly explains the two key indicators of social sustainability, which are social cohesion and social justice; and then proceeds to introduce social capital, as well as the adaptation of social capital on other research fields, based on which, proposing the adaptation of social capital approach on the field of urban conservation in order to promote the sustainable development, especially in the social dimension.In addition, Chapter Three also provides a brief study on the specific social relationship pattern in traditional Chinese society, which is "CHA-XU-GE-JU"(差序格局) , and the core social value which is attached on the traditional Chinese residential neighborhood——HE(和), namely social cohesion is explored. Furthermore the five key indicators of HE(和) which are the model of "Benevolence-Righteousness-Propriety-Wisdom-Faithfulness"(仁义礼智信)are also discussed for further adaptation of social capital survey in the selected cases.Chapter Four proposes the research methodology, which aims to test the hypothesis via mixed method including both qualitative and quantitative methods. Social capital survey and social network analysis under the decision-making framework of path dependency are employed.Chapter Five and Six utilize two selected cases to reveal the conservation process from two different cities of China, and to inductively understand the role of social capital in urban conservation. Chapter Seven presents the data analysis of the social capital survey and social network analysis based on the above two cases. Chapter Eight presents the discussion of the findings, in light of which, this chapter goes further to discussthe generalization of this research; in the end it proceeds to summarize the conclusions of this research, its limitations, contributions, as well as the implicationsfor future studies.

Chapter 2   Literature Review of Urban Conservation

2.1   Definition of urban conservation

2.1.1   Urban conservation, preservation and restoration

The definition of urban conservation has evolved architectural preservation and restoration. To understand conservation, it is necessary to distinguish between the following terms——restoration, preservation and conservation——as there are ambiguous uses of these terms (Rodwell 2007).[①]

In Burra Charter restoration is defined as "returning the existing fabric of a place to a known earlier state by removing accretions or by reassembling existing components without the introduction of new material"(ICOMOS 1999), and preservation is defined as "maintaining the fabric of a place in its existing state and retarding deterioration"(ICOMOS 1999). By applying physically curatorial and architectural methods, these two definitions address the importance of the historical origins and material authenticity[②] of historical buildings based on their previous or current states, while it lacks the consideration of buildings contemporary adaptations and future uses.

During the early development stage of the concept "conservation", it was argued that historical remains should be restored to an earlier stage or preserved in their original condition in terms of building materials, façade, styles and so on (Ruskin 1884), which is subsequently criticized to be "extreme and impracticable" (Larkham 1996).

Along with the conservation practices and the development of meanings of heritage , an alterna-

---

[①]   For example,in Athens Charter the fundamental principles of restoration and protection of monuments like codification on a basis of international collaboration was established Charter, A. (1931). The Athens Charter for the Restoration of Historic Monuments. the 1st Int. Congress of Architects and Technicians of Historic Monuments, Athens, Greece.. In Venice Charter, the significance of permanent preservation of monuments was emphasized Charter, V. (1964). International charter for the conservation and restoration of monuments and sites. IInd International Congress of Architects and Technicians of Historic Monuments, Venice. Nevertheless, in these two charters the following terms——"restoration", "preservation", and "conservation" are not clarified.

[②]   The term "authenticity" wasfirstly conceived in Charter of VeniceICOMOS (1964). "Charter of Venice." ICOMOS. and was defined by the International Center for Conservation in Rome as "materially original or genuine as it was constructed and as it has aged and weathered in time" Feilden, B. M. and J. Jokilehto (1993). Management guidelines for world cultural heritage sites, ICCROM. In the Nara Document on AuthenticityICOMOS (1994). Nara Document on Authenticity. Nara Conference on Authenticity, the linking of the diversity of authenticity judgments with the variety of sources was identified and elaborated.

tive view argued that conservation should not only include preservation and restoration, but also involve revitalization of historical places (Buchanan 1968), as Larkham(1996) appealed that *"changes should be permitted to allow a new function to inject life into the building and its surrounding, even at the cost of some alterations."* This assertion highlights the adaptive reuses to be added to the domain of conservation. It also calls for a divergence of conservation discourse from the absolute physical dimension: "As a field, we have come to recognize that conservation cannot unify or advance with any real innovation or vision if we continue to concentratethe bulk of conservation discourse on issues of physical condition"(Avrami, Mason et al. 2000).

Given that the changes in other arena such as social change and cultural change are manifested in the field of conservation, Tiesdell, et al. (1996) further clarified the definitions for preservation and conservation as such: *"Preservation is concerned with limiting change, conservation is about the inevitability of change and the management of that change."* It is now widely accepted that the conservation discourse has swung from physical preservation to a commonly accepted perspective, which addresses that conservation should adapt to change and manage change(Larkham 1996, Tiesdell, Oc et al. 1996, Avrami, Mason et al. 2000, Hobson 2004, Rodwell 2007, Madgin 2009). This evolution extends the scope of conservation from architectural practices on physical remains to the management of historical places, albeit the core of conservation has not been emphasized. More importantly, the need to contextualize the term "change" has been highlighted. Firstly, the categories of "change" go beyond the physical dimension to a wider sphere which includes both internal driving forces such as the continuously evolving heritage values and place, and external driving forces such as constantly altering contexts. Secondly, the change in various dimensions will influence conservation decision-making, "The decisions about what to conserve andhow to conserve are largely defined by cultural contexts, societal trends, political and economic forces——which themselves continue to change"(Avrami, Mason et al. 2000). In light of this, conservation needs to be contextualized in order to understand the change.

Meanwhile, influenced by cultural relativism, Burra Charter gives rise to the exploration of heritage value in the cultural dimension, and cultural significance of a place is introduced to redefine heritage value(Wells 2007). In Burra Charter, conservation is further defined as *"all the processes of looking after a place so as to retain its cultural significance"* (ICOMOS 1999), with the cultural significance[①] meaning——*"aesthetic, historic, scientific, social or spiritual*

[①] According to Burra Charter, "cultural significance is embodied in the place itself, its fabric, setting, use,associations, meanings, records, related places and related objects"; and "places may have a range of values for different individuals or groups" ICOMOS (1999). The Burra Charter: The Australia ICOMOS charter for places of cultural significance 1999: with associated guidelines and code on the ethics of co-existence, Australia ICOMOS.

value for past, present or future generations" (ibid.).

In the first place, this definition has highlighted the position of cultural significanceas a set of heritage values in the field of heritage conservation, which could be articulated as such: the core objective of conservation is to retain cultural significance of heritage, namely heritage values of a place. As heritage value embraces plural interpretations, it is necessary for conservation to clarify "what", "why", by "whom", and for "whom" heritage is cherished, as well as "how" to conserve. As such, conservation is a social performance, intertwined with diverse social groups and individuals.

In the second place, in order to accomplish this goal, conservation "processes" are developed to manage the change. The term "process" could be explored in the following aspects: 1) The identification of cultural significance is a cultural process, with the constantly evolving understandings of heritage values ; 2) The conservation planning, from the identification, assessment, and selection of heritage value to the development of management tools, adaptation mechanisms and the implementation of interventions, requires a comprehensive planning process; 3) The integration of diverse social groups , individuals, together with the various practices embedded in specific times, contexts and space, which allow the meaning of heritage to be redefined and recreated constantly in response to certain social trends. As such, heritage conservation could be further understood as "a bundle of highly politicized socialprocesses, intertwined with myriad other economic,political, and cultural processes"(Avrami, Mason et al. 2000).

Building on the above explorations, urban conservation should be recognized as a collection of heritage value centered place management processes to manage change.The core role of the discipline of conservation remains to retain the values of imbued in heritage, the scope expands from architectural preservation to the place management processes, and the understanding of change transcends the physical dimension towards the multiple dimensions.

On one hand, different ways to value heritage accordingly will engender different conservation approaches. And as the understanding of heritage value evolves, the discipline of urban conservation needs to respond accordingly, and adapt to new agendas with new approaches; for instance, the evolution of conservation doctrine related to historic areas and the adaptation approaches to the heritage environment conservation management.

On the other hand, it also highlights the need to reconcile the potential impacts of emerging agendas on heritage values to the orthodox of urban conservation planning which established comprehensive principles, regulations and so on. One example could be the postmodern perspective of heritage value selection which favors pluralism and diversity of individuals and might lead to the reduction of orthodox conservation values such as authenticity and the fragmentation of existing modern conservation decision-making framework (Pendlebury 2008).

2.1.2 Heritage and heritage value

Heritage has been a widely employed term in the fields of conservation, tourism, cultural industry, etc. Heritage is defined by UNESCO as *"our legacy from the past, what we live with today, and what we pass on to future generations"*.[①] In 1972, the UNESCO Convention on the protection of the World Cultural and Natural Heritage (Paris)presented the idea of cultural heritage and natural heritage and identified monuments, groups of buildings and sites as three categories of cultural heritage(UNESCO 1972). The definition of heritage by UNESCO in 1972 is comparatively broad, and the subsequent international charters and convention documents has explicitly elaborated on heritage in many aspects, among which, Nara Document on authenticity (ICOMOS 1994) which extends the meaning of heritage to the diversity of authenticity judgments and addresses the diversity of cultural values, the 2003 Convention for the safeguarding of the intangible cultural heritage(UNESCO 2003) which explores the intangible heritage value particularly, together withthe Quebec Declaration on the preservation on the spirit of place which tackles the issue of living heritage in the continuous process of spiritual reconstruction based on various contexts and inclusive values of social groups (ICOMOS 2008), provide new avenues on the understanding of the scope of heritage.[②]

In 1980s the significance of heritage selection was acknowledged as to promote heritage related activities like tourism, as Larkham(1996) argues that heritage is a process, which aims to gain popular consumption mainly through selecting and presenting history and place. In this process, the selection of heritage, which means the selection of heritage value to be presented is a vital part(Ashworth 1991). Nevertheless,the selectivity of heritage value is recurrently questioned.

In fact as the selections of meaning and significance are frequently confronted with complicated cultural, political, social, and economic contexts, it is challenging to be sufficiently inclusive for heritage selection. For example, as it has been widely argued, the local community is usually excluded in the selection process (Larkham 1996). Similarly, Rodwell (2007) criticized that heritage is *"perceived to be divorced from individual and community life today"*. He goes further to

---

① Website of UNESCO World Heritage Centre——World Heritage , http://whc.unesco.org/en/about/.
② The international charters and convention documents cover a wide scope of heritage including buildings, urban areas, towns, gardens, wall paintings, material and structures, underwater cultural heritage, archeological heritage, industrial heritage, cultural routes, spirit of place, and so on and so forth. (source:http://www.icomos.org/en/charters-and-texts;http://whc.unesco.org/en/about/).

argue that this inadequate understanding of heritage has been finally attributed to the limitations of architectural conservation objectives, which are primarily the endeavors to preserve the historical remains and to motivate the heritage industry (Rodwell 2007).

Nevertheless, the understanding of heritage value has not ceased to evolve.Starting from the 1980s, several trends indicating how heritage adapts to the postmodern critiques on the discourse of urban heritage conservation have gradually come to the surface. Firstly, in terms of heritage value identification,the subjective values of heritage and the living performance, such as identities, experience, memories, meanings and so on and so forth, which are attached to the physical fabrics have been recognized and advocated in the discourse of urban conservation(Smith 2006, Wells 2007, ICOMOS 2008, Pendlebury 2008, Wells 2010, Wells and Baldwin 2012). Secondly, in terms of value identification approach, a shift emerges from the experts-led "fabric centered" conservation approach to the "values centered" stakeholders inclusive approach (Mason 2008, Wells and Baldwin 2012).Thirdly, heritage has been understood as "a cultural and social process", which through conservation management on the legacy of past, "create the ways to understand and engage with the present" (Smith 2006).

Through the examination of the heritage conservation charters[1] under a framework of discourse theory, Wells(2007) identified that the value of heritage has shifted from a "positivist truth based on substantiation of material fetish" as an absolute standard, together with preservation and restoration as the "hegemonic discourses" towards "the relative truths" based on cultural pluralism and the diversity of contexts. He goes further to argue that the interpretation of heritage value and the construction of heritage significance "must be based on complex and interrelated meanings bound into a cultural milieu" (ibid.).Hence rather than from a positivist perspective, through the lens of cultural relativism which looks into the specific contexts of heritage, individuals and various social groups, the pluralism of cultural significance as heritage value is advocated.

To further understand the cultural significance of urban heritage, assessment tools[2] have been developed, and values are examined by many categories such as "fundamental and incidental values", "emic and etic meanings", "valuing and valorizing", and more. Nevertheless, values are complex and fluid, just as Pendlebury argues: "The nature of heritage isthat it is socially

---

[1] They are namely the Athens Charter , the Venice Charter , and the Burra Charter.
[2] For example, The Getty Conservation Institute introduced a "provisional typology of heritage value" to examine the cultural significance of heritage based on the category of social-cultural value and economic value. Mason, R. (2002). "Assessing values in conservation planning: methodological issues and choices." Assessing the values of cultural heritage: 5-30.

constructed; value is never an intrinsic quality but is externallyimposed according to culturally and historically specific frameworks. Thesemay be culturally, or temporally, collective or may be very personal; we eachhave our own value frameworks."A majority of value identification approaches conduct "reductionist","one-time", and "limited"approaches to examine cultural significance, thus failing to understand and reveal the richness of heritage values(Avrami, Mason et al. 2000, Pendlebury 2008).

A further reflection herein lies in the fact that both the conservation doctrine and the methodology should adapt to the introduction of cultural significance as heritage values and to the integration of diverse contexts(Mason 2002, Wells 2007):"despite emerging policies that promote value-driven planning for conservation management, there is a limited body of knowledge regarding how conservation functions insociety——and specifically regarding how cultural significance might best be assessed and reassessed aspart of a public andenduring conservation process." The need to integrate value assessment into material fabric based conservation calls for the knowledge of social sciencewhich could facilitate the assessment of value, identification of interest groups and individuals——stakeholders, the involvement of stakeholders in the decision-making process under various contexts, etc.(Avrami, Mason et al. 2000).

2.2 The evolution of urban conservation doctrine in Europe (1800s—2000s)

2.2.1 The emergence of modern "Architectural Conservation" (1800s—1930s)

It is commonly accepted that the development of modern architectural conservation stems from the turn of the eighteenth century to the nineteenth century (Larkham 1996, Hobson 2004, Rodwell 2007). The external driving factors are: 1) the development of science and technology in the eighteenth century——Age of Reason in Europe evoked the changes in many fields such as culture, belief, social study, modern archaeology and arts; 2) the progress in modern archaeology and art history inspired the idea of authenticity and a growing interest of the antiquity. Meanwhile, the changing attitudes towards antiquity evoked the study of traditional architectural style, construction technology, building materials and old craft skills (Rodwell 2007), and it also influenced the preservation attitude in architecture, which is to honor the historic remains by keeping them untouched(Ruskin 1884); 3)The rise of picturesque aesthetic gradually prevailed in the fields of architecture and art, which finally led to the Picturesque Movement , which preferred the natural and nostalgic aesthetic features.

The internal impetuses are: 1) the rethinking of mass production, which is attributed to have caused the separation of arts and crafts, and led to many social and environmental problems

(Kennet 1972, Hobson 2004); 2) and the rising of social elites such as Ruskin and Morris, who realized the value of historical buildings and relics and became leading proponents of preservation and anti-restoration practices (Larkham 1996, Pendlebury 2008). In addition, the establishment of the Society for the Protection ofAncient Buildings (SPAB) with its famous manifesto which marked the rise of modern conservation campaign which introduced the key concept of "authenticity based on material fabric"(Pendlebury 2008) and remains influential in contemporary"material fetish" preoccupied daily conservation practices(Wells 2007).

Influenced by above factors, in the late nineteenth century, a group of countries, namely Britain, the Netherlands, France and Germany initiated "the first period of conservation legislation" and started the listing of important monuments (Larkham 1996), which according to Tiesdell et al.(1996), can be identified as the first wave of modern conservation movement——the architectural conservation. In physical aspects, it seeks the authenticity of relics, and takes a *curatorial approach* to protect historic remains as monuments. As it is mainly driven and dominated by social elites such as architects, archaeologists and other specialists, who have displayed specific architectural and historic interests and preferred the isolated antique monuments instead of the urban historic buildings, rigid protections were given priorities over other requirements, such as working class's needs (Larkham 1996).

2.2.2 The rise of "area-based conservation planning" and the struggles to balance conservation and development (1940s—1960s)

During the after-war clearance and rebuilding period from 1940s and 1970s, property developers adopted modern architecture and planning methods, such us in the British Newtown Movement (Alexander 2009). The reconstruction planning and slum clearance policy together with the subsequent road building schemes in postwar decades dramatically changed the social, cultural and built environment, which lead to many consequences such as a hasty gap between past and present in many cities and a growing opposition of the demolition policy (Pendlebury 2008, Alexander 2009). The period witnessed the introduction of anarea-based "comprehensive conservation system" through which the role of the government has transformed from "a bystander" to "a provider and implementer"(Pendlebury 2008).

After two World Wars, strengthened by the growing awareness of preservation in the public sector, such as the 1967 Civic Amenity movement in the UK, second major phase of legislation emerged.This extended the scope of conservation from individual historical preservation practices to public movement (Cullingworth 1999), and encouraged innovative ideas (Hobson 2004), a few examples are: the idea of the "protected perimeters" and the protection of "the groups of buildings" in France, and "the conservation of the entire areas" in UK(Larkham 1996), which embarked on "*an important shift in emphasis away from preservation the individual historical buildings towards the conservation of coherent areas of townscapes*" (Cullingworth 1999). Accordingly, during this period, the principle toward historic remains shifted from rigid obedience to the past towards embracing both the old and the new, as it has been stated in Geddes'(1949) *Cities in Evolution* and Gullen's (1961) coherent townscape.

Based on the above contexts, during this period, most European countries witnessed the significant emergence of "area—based" conservation movements, including the conservation of township and street patterns. Tiesdell et al.(1996) identifies the stage of area—based conservation as the second wave of conservation.

The approach of designation was employed to identify conservation areas.[1] Nevertheless, it is usually questioned because it was solely based on "special architecture or historic interest" (Civic Amenities Act, 1967), and because the value judgment of designation was largely limited to architectural and historic sectors, hence ignoring the social and economic factors. Furthermore, in terms of the designation method, it is also criticized for being preservation based and for highly relying on technical planning mechanisms, such as land use zoning and transportation. For instance, the strict development control in conservation area restricted the functions, uses and everyday life of some ordinary buildings within the preservation zone, which not only caused serious economic and social problems such as livelihood loss, but also were prone to change a dynamic mix-used place into a museum of tourist spots(Burtenshaw, Bateman et al. 1991, Hobson 2004).

Accordingly, the functional considerations of areas and the adaptive reuse of aging buildings were added to the previously simple physical-conservation based measures, and a broad discussion emerged on conservation addressing the urge to create a distinctive approach from the "*museological*" preservation approach(Hobson 2004, Rodwell 2007). For instance, Gustavo Giovannoni (1873—1947) proposed the idea of "*living conservation*" and developed the idea of "*urban heritage*", which addressed the importance of function of historic buildings in an urban context (quoted from Rodwell 2007).

In the UK, pilot studies for the historic town master plan were conducted in four historic towns, namely Bath, Chester, Chichester, and York during 1966 to 1968. Aiming to reconcile the pressure raised by needs of modern functions such as traffic versus pedestrians,these historic town plans "all advocated a balance between the conservation of historic character and the continuing

---

[1] Conservation areas are defined as "*areas of special architectural or historic interest the character of which it is desirable to preserve or enhance*"UNESCO (1967). "Civic Amenities Act ".

evolution of the living city, albeit the balance suggested varied significantly amongst the various plans"(Pendlebury 2008).①

During the shift of the conservation approach from architectural preservation to the conservation of areas, the conservation ideology has evolved from mere historical preservation to a comprehensive urban enhancement(Ashworth and Tunbridge 1990). This evolution in conservation perspectives is revealed in the charters and declarations of ICOMOS and UNESCO during the 1960s till 1970s. Although the Venice Charter(ICOMOS 1964) declared the historic monuments as major conserved objects and the authenticity of materials and documents as a basic concern, it also gives attention to the importance of safeguarding the historical sites of monuments. In 1972, monuments, groups of buildings and sites were identified as three categories of cultural heritage by UNESCO(UNESCO 1972). Shortly afterwards, the value of historic areas and the role of contemporary condition in the conservation of historic area were addressed in the Warsaw-Nairobi Recommendation, which proposed "international cooperation" and the conservation measures integrated in "legal, administrative, technical, economic, and social aspects"(UNESCO 1976).

2.2.3 The "policy—initiated Conservation" (1970s—)

In the UK, the relationship between conservation and urban planning began to change dramatically in the late 1960s: "Conservation ceased to be a backdrop to a process of modern planning;rather it began to assume the foreground and become the starting point. Itstarted to encompass the whole quality of the environment, including the insertion of new buildings. It gradually became the driver in the managementof change"(Pendlebury 2008). Yet shortly after this period the balance was broken by the economic situation in 1970s and the rational choice based modern planning approaches such as "compulsory purchase" and large scale redevelopment plan. Especially in the traditional residential area, theirreversible displacement method with the dispersal of lower social class is often accused of destroying the "sense of place" of the historic area and of causing feeling of "uncertainty" (Tiesdell, Oc et al. 1996, Pendlebury 2008),thus engendering the recurrent protests of communities against the demolition and redevelopment such as the "NIMBY"movements which have been occurringall over the world ever since late 1970s, arousing new agendas to the domain of urban conservation such as public participation, community development, social conservation, the integral selection of heritage value, and so on

---

① The plans recognized the importance of mixed use neighborhood and the need to introduce new functions for the impacted historic buildings and advocated that it is only "through comprehensive planning" with "balanced approaches" that the evolution of historic cities can be reconciled in modern circumstance. Particularly in Esher plan of York, a progressive planning facilitated with approaches such as small scale redevelopment, intensification of residential land use, and the control of traffic was recognized as a benchmark of "a more inclusive and embracing conservation of place" and has been influential in historic town planning Pendlebury, J. (2008). Conservation in the Age of Consensus, Routledge..

and so forth. Just as Pendlebury(2008)points it out:"planned redevelopment fired the growth of the conservation movement,but conservationists sought not the abolition of planning but to take control within it", conservation remains as the practices of modern planning in many aspects, and hence has kept on evolving through continuous adaptation to the emerging agendas and to the specific local contexts in order to manage the process of change.

Given the economic recession in the 1970s as a significant influential factor, in UK the 1980s' legislation retrenched the official support for conservation (Larkham 1996). Instead of official financial support, the economics of conservation tends to extensively rely on the economics of urban development (Lichfield 1988, Pendlebury 2008). To gain economic growth, conservation was frequently used as a governmental tool in dealing with many other areas of urban development, such as housing and economic regeneration (Lichfield 1988, Urry 1995, Larkham 1996, Tiesdell, Oc et al. 1996, Pendlebury 2008).

During the last two decades of the 20th century, various conservation practices have been conducted, which focused on the attraction of investment as well as the economic growth of the area in order to revitalize historic urban quarters such as the SOHO in New York and the Jewellery Quarter in Birmingham. Based on such conservation practices, Tiesdell et.al.(1996) identifies an emerging common feature of this conservation stage as *"the revitalization of the protected historic urban areas and quarters through growth management"*, which indicates the rise of the third wave of conservation: Policy-Initiated conservation. Compared with the previous preservation-oriented approach, which focused on the historic interest of *'the legacy of past'*, the policy-initiated conservation highlights the concern of the potential development of heritage. In terms of conservation measures, policy-initiated conservation targets to manage both the physical change and the inevitable economic change.

According to Lichfield (1988), the policy-initiated urban revitalization aims to remedy all kinds of obsolescence in order to extend the life circle of built environment, which will in turn motivate potential economic activities. According to Tiesdell (1996), specific strategies are conducted to achieve physical and economic revitalization. In physical revitalization, a short-term preservation-led physical revitalization is usually adopted to reduce the physical obsolescence and to sustain people's confidence. To promote economic growth, three approaches are widely used(Tiesdell, Oc et al. 1996). The first one is functional restructuring which replaces the existing activities and functions by new uses and activities. The second one is functional regeneration which maintains and enhances the existing functions and improves the competitiveness of the area's existing functions and uses. The third one is functional diversification which injects in some new feasible uses to the historic urban area. This period saw the rise of the heritage industry which commodified the concept of heritage. Hence practices of urban heritage conservation have been woven into broader urban agendas such as housing, real estate and tourism, which in turn engendered new phenomena such as gentrification during the process.

The economic and physical revitalization has provoked concerns on the negative social consequences of gentrification. As can be seen from many practices, the functional restructuring has changed the entire economic base and function of places(Lichfield 1988). As its aim focuses on land value increase and rent generators, as well as on attracting higher economically valued uses, it also replaces the activities of the historic urban area and users(Lichfield 1988, Tiesdell, Oc et al. 1996). Since urban functions, people's activities and their indigenous local social network were replaced, the social values which are all deeply embedded in close relationship between neighborhoods, such as the sense of trust and security, and the collective memory, etc. have been dramatically damaged.

2.2.4 New agendas of urban conservation (1980s—2000s)

Starting from the 1980s, in conjunction with the urban conservation practices a shift of defining heritage value emerged from a singular standard of value based on material fabric towards heterogeneous values based on postmodern plurality, cultural relativism, thus providing new avenues for understanding the concept of heritage and developing fluid and inclusive approaches for heritage conservation (Wells 2007, Pendlebury 2008).

As the impacts of urban conservation practices become apparent, for instance, the demolition projects lead to recurrent resistance, the role of urban conservation on social inclusion are receiving growing attention, thus urging urban conservation planning to adapt to new agendas related to the social role of urban conservation such as plurality and diversity, community identity enhancement, daily life heritage, constructing social inclusion, and public value. Measures such as "characterization" of historic environment, and "catch-all"which incorporate broader values have been implemented in the UK, however, the translation of pluralism into the orthodox conservation planning might lead to the fragmentation of the established modern conservation(Pendlebury 2008). In addition, it is argued that in general it lacks both doctrine decrees and methodological means to facilitate the decision-making frameworks of conservation practices(Wells 2007, Pendlebury 2008).

In the advent of an age of consensus[①] on urban conservation, potential challenges have now

---

[①] Pendlebury recognized that urban conservation is now in the "age of consensus" when it is agreed worldwide that heritage protection is important (ibid.).

come to the surface in the process of adaptation: 1) firstly, in terms of value definition approaches, the question on how to reconcile the calls for pluralistic values which favor individual choices with the framework of value selection of orthodox conservation planning which is rational choice based, professionals defined, and authenticity of material fabric centered;2) secondly, in terms of conservation planning management, the challenge to adapt these agendas into the legislation framework and interventions such as strategies, regulations, and principles etc., hence embracing an inclusive decision making process.

2.3 Urban conservation practices in global perspective (1970s—2000s)

The development of urban conservation can be mainly concluded in three stages, namely architectural preservation(1800s—1930s), area-based conservation(1940s—1970s), and policy-initiated conservation(1980s—2000s)(Tiesdell, Oc et al. 1996, Pendlebury 2008). Especially in the third stage, urban conservation has been largely involved into urban development and has adopted many approaches according to different urban development strategies. The followings section will introduce some specific measures adopted in urban conservation practices all over the world.

2.3.1 Approaches adopted for urban conservation in historic areas in Europe and the United States

1) *Top down approaches*:
- The policy-initiated revitalization in historic areas

According to Tiesdell(1996) and Roodhouse (2010), the following four measures were adapted into the policy-initiated urban revitalization in urban historic areas to merit different kinds of obsolescence and to facilitate urban development in other aspects such as tourism and housing:

- Tourism-led revitalization through a process of place-marketing

Via a place marketing process (Kearns and Philo 1993), tourism is usually accompanied by the idea of heritage to revitalize decayed historic urban quarters such as *Castlefield Quarter, in Manchester* (Tiesdell, Oc et al. 1996), aiming to promote the area's economic growth, to increase the use of its historical resources, and to reduce the image obsolescence.

- Housing-led revitalization

By employing adaptive reuse of urban historic quarters, this measure aims to revitalize dilapidated quarters by attracting residential use back to city centers, given that old quarters usually occupy good locations, and have attractive historic characters and convenient accesses to infrastructure and facilities, such as public transportation and playgrounds. In addition, once the adaptive reuse of old buildings provides residential supply, economic and social activities will return, hence the further declines of old urban quarters could be prevented. Furthermore, by increasing city center living, the mix-used vitality of old quarters can be enhanced, as it can be seen in the cases of *Marais quarters in Paris* (Rodwell 2007) *and Glasgow's Merchant City in UK* (Tiesdell, Oc et al. 1996). Although housing-led revitalization strategy aims to provoke a regeneration process of the larger scale, the historic area could also be threatened by gentrification(Larkham 1996, Tiesdell, Oc et al. 1996, Cullingworth 1999).

- Cultural-led revitalization through process of place-making

Through the enhancement of the quarter's cultural atmosphere and unique characters in aspects of architecture, history, and local everyday activities, etc., the cultural areas are attractive to both tourists and local people(Murray 2001). For example, in *Temple Bar of Dublin* (Tiesdell, Oc et al. 1996) and *Triangle Below Canal Street*, New York(Tiesdell, Oc et al. 1996), through the culture led place making process, both local cultural production and cultural consumptions from both tourists and locals have been stimulated and promoted, such as accommodation, catering business and traditional retails of local handcrafts (Tiesdell, Oc et al. 1996).

- Creative industry-led revitalization

Inspired by Soho in New York, a trend called "creative dismissions" emerged—— where the decayed urban places have been transformed into creative cultural hubs(Landry 2008, Roodhouse 2010). According to Roodhouse(2010), creative dismissions means " *the conversion of a facility previously used for the non-cultural purposes into a cultural and creative functional centre.*" These areas are usually places with previously low real estate price. Artists came in and run their design workshops, art studios, and art galleries at low rent; later followed the entertainment sectors such as bars, coffee shops, and fashion shops. As a result, the decayed areas have been revitalized not only in the daytime, but also in the night time. While "creative dismissions" works as a catalyst for the "cultural transformation of urban space" (Roodhouse 2010), it still faces both the risk of massive gentrification and the fleeing away of creative people if the rental prices are rising continuously (Tiesdell, Oc et al. 1996, Roodhouse 2010).

- Legislative provisions and intervention: policy, fund, and tax measures

Starting from Civic Amenities Act, legislative support in UK occurred mainly during the period of 1967—1979 (Larkham 1996). As the reuse of inner city old buildings was initiated by government, legislative support and intervention became an important policy-incorporated approach to secure the implementation of rehabilitation (Larkham 1996, Pendlebury 2008).

In the U.S., accompanied by the changing contexts of environment movement in 1960s——social demographic change, working couple life-style, energy concerns, property price and critics of inner city decay and urban sprawl[①]——the rehabilitation of inner city old buildings became a major objective as an incentive to bring people "back to the city" (Collins 1980).The 1966 National Historic Preservation Act (US Congress 1966) addressed the preservation of historic buildings and urban rehabilitation, which *"offered a foundation for a form of urban conservation which is now the central aspect of historic preservation practice"* (Collins 1980).

To stimulate the movement of historic district rehabilitation and attract people "back to the city", a series of supportive legislative provisions were issued. Firstly, a federal funds system was connected with the National Registration system, which advocates the registration of historic buildings. Secondly, the Tax Reform Act of 1976 and 1978 reduced the financial support of new construction development and offers tax credits for rehabilitation (Lee 1977, Collins 1980, Wells 2010). Thirdly, *"Internal Revenue Service allows deductions for charitable contributions for easement or property rights conferred for historic preservation purposes"*(Collins 1980). The legislative support and intervention mostly through fund and tax benefit largely contributed to the movement of the inner city rehabilitation and to the returning of people to the city.As Wells argues (Wells 2010), the Main Street Program has triggered the phenomenon of "Spontaneous Fantasy" which evokes creative experiences of spaces and adds new collective memories to the decayed ones. Furthermore, Public Buildings Cooperative Use Act of 1976 (US Congress,1976) issued the reuse of historic buildings, and many historic buildings, for example SOHO in New York have been converted into housing apartments and commercial uses, which provided more economic opportunities for historic buildings.

As house owners seek the higher rental profit, historic buildings are revitalized to attract high end activities and upper class. Accompanied with the physical and economic revitalization, many historic areas experienced a process of economic restructuring and social replacement: gentrification (Collins 1980, Tiesdell, Oc et al. 1996), which is a well-known process in the inner city rehabilitation in the United States, and starts with housing-led revitalization and later occurs in other sectors such as tourism (Tiesdell, Oc et al. 1996).

2) *The bottom-up community involvement approach*
Ever since 1990s, as it can be seen in many community-based projects, through community

---

① Before the 1966 National Historic Preservation Act was issued, the urban development in the United States favored the suburban development mode and the renewal clearance of old buildings, andconservation practices had been limited on the architectural preservation of monuments and buildings with special historic and architectural interests. Collins, R. C. (1980). "Changing views on historical conservation in cities." The ANNALS of the American Academy of Political and Social Science 451(1): 86-97..

based measures such as heritage education, environmental campaign, revival of traditional crafts and techniques, and school projects, the bottom-up local "community involvement" approach has been adopted in the European cities (Rodwell 2007), encourages community participation in the decision making process of both tangible and intangible heritage movements, andenhances cooperation between community and institutions.

- Main Street Movement in US: community based downtown revitalization

The Main Street Project was launched by The National Trust for Historic Preservation in 1977 to promote downtown revitalization (Nichols 1996, Robertson 2004, Wells 2010). By applying the "Four Point Approach"——: *"organization, promotion, design and economic restructuring"*[①], this project used a community based development approach, which aims to revitalize the traditional community in downtown areas through management and the coordination (Nichols 1996, Robertson 2004). Based on the participation of stakeholders, an *organization* from Main Street produced a vision and a development plan. *Promotion* usually includes image enhancement, marketing retail and festivals activities (Nichols 1996, Robertson 2004), which contribute to physical and economic revitalization. *Design* aims to tackle with the preservation and renewal of old buildings and spaces. *Economic restructuring* is vital to the Main Street Program as it enhances regular business and injects new economic activity functions to existing communities. The Main Street approach highlights the cooperation between public and private sectors and the management based community development program(Wells 2010). This approach adopted "revitalization culture" values, even though deviated from "the grain of conservation doctrine", it spontaneously generated a sense of "authenticity" and "attachment" of the pluralistic culture, through the social and cultural process of value construction (Wells 2010). This method has been widely adopted by city planners in many cities (e.g. St. Charles, Illinois; and Cushing, Oklahoma, etc.)and into many community-development based downtown revitalization practices(Nichols 1996, Robertson 2004).

3) Incorporated conservation management to achieve rehabilitation of historic areas
European cities have been actively promoting the concept of "integrated conservation". The campaign themed as *Europe,A Common Heritage* was launched in 1999 which has promoted largely the integration of heritage conservation in terms of experience exchange and cooperation between historic towns, collaborations with organizations, public education and participation, partnership, conservation management and policy making, and so on (Pickard 2001).

---

① National Main Street Center, http://www.preservationnation.org/main-street/about-main-street/the-approach/.

Based on the case studies of ten European cities including Bruges, Dublin, Grainger Town, etc. conducted by many other scholars, Pickard(2001) indicates that European cities have a tendency to adopt an approach of incorporated management of historic areas, *"rehabilitation and the role of management strategies in historic centers has become the subject of debate and analysis by the Council of Europe's technical co-operation and consultancy program in the field of cultural heritage"*. This incorporated management approach includes three major themes, namely *"Urban Values", "Politicaland Institutional Framework"*, and *"Intervention Methods"* in the management process to obtain rehabilitation (Pickard 2001).

Urban values are examined through a category of values system such as "identity and diversity", "respect for the morphology and typology", "the importance of public areas", "the perception of architecture", "social diversity" and "functional diversity" (Pickard 2001).

As conservation management needs an integrated institutional framework and political supports, the theme *"Political and institutional framework"* examines not only the political commitment at the national level in terms of financial support and legislative provisions, but also the associated cooperation and participation of public and private sectors from the local level. Four aspects are discussed: "political commitment", "rehabilitation as a component of official housing policy", "an integrated approach to planning", and "striking a balance between public and private sectors"(Ibid.).

Management and *intervention measures* are required for the successful implementation of projects. Various measures are applied in the project management process, including : to "keep the project on a manageable scale", "interpretation as a factor of appropriation", "democratic management" , "decentralizing intervention teams", to "incorporate with top-down and bottom-up strategy", "environmental and tourism management" (Pickard 2001).

2.3.2 Approaches adopted for urban conservation of historic areas in Singapore and Japan

Asian countries present great diversity in terms of cultural backgrounds, development levels, social structures and political systems. In terms of urban conservation, Asian cities exhibit various approaches based on diverse social, economic, and political contexts. Within considerable diversity of cultural and social contexts, many practicable conservation measures have been developed. For example, in Taiwan, the adaptive reuses of colonial cultural heritage and heritage tourism (Blundell 1992, Logan and Reeves 2008, Cheng and Fu 2010), community participation and collective memory (Hisa 1998, Hung-Jen and Waley 2006), cultural regeneration and community mobility(Lin and Hsing 2009)have been widely discussed and practiced.

Aside from the diverse contexts, most of Asian cities bear certain similarities. Firstly, most Asian countries are categorized as developing countries, with many of them experiencing rapid urbanization. Secondly, given the long history of civilization, most of Asian cities have an abundance of heritage resource. In addition, Asian cities accommodate a large population compared with European and American cities. Accordingly, a relative higher density is another common characteristic of Asian cities. Considering that all shared common features, some urban conservation approaches from the most developed Asian countries, namely Singapore and Japan might bring light into other Asian cities' approaches.

1) *Approaches adopted for urban conservation in historic areas in Singapore*
Although the attempt to preserve part of Singapore's heritage was firstly mentioned in the United Nations report, Growth and Urban Renewal in Singapore in 1963 (Ghosh, Gupta et al. 1996), Singapore did not start the urban conservation practices until 1970s when eight national monuments were listed (in 1973).[①] The urban renewal strategy in 1960s and 1970s has successfully improved the situation of Singapore's economy, public housing, technology and infrastructure, which however, together with slum clearance, resettlement of residents, and the decentralization strategy of the traditional businesses and warehouse industries in waterfront areas, has caused further obsolesces of old quarters such as Niucheshui (Chinatown area),Boat Quay and Clark Quay in the 1970s(Lee 1996, Chang and Yeoh 1999, Heng and Chan 2000, Zhu, Sim et al. 2006, Heng and Low 2009). In addition, the calls for heritage tourism, together with the growing desires of identity and historical continuity also added to government's awareness of urban conservation (Teo and Huang 1995, Yeoh and Huang 1996, Chang and Yeoh 1999, Heng 1999).

Upon the renovation of several "state-owned shop houses"[②] in the 1980s, In 1986 the Urban Redevelopment Authority (URA) of Singapore released The Conservation Master Plan, with a changing focus from monuments to historic areas. In 1989, 10 historic areas were firstly designated by URA.[③] The urban conservation designation system has several degrees. Up to now,

---

① "A brief history of conservation in Singapore", Source: http://www.ura.gov.sg/uol/conservation/vision-and-principles/brief-history.aspx.
② In the Central Area of Singapore, the main urban form is composed by traditional shophouses, which is the traditional architectural type of mix-used, low rise row-houses with each consisting of a narrow front with the public five-foot ways, and are deep in length with an interiorair well or courtyard. Originally, the ground floor was used for retailers, wholesalers, and workshopswhile the upper floor was residential. Wing, H. C. and S. L. Lee (1980). "The characteristics and locational patterns of wholesale and service trades in the central area of Singapore." Singapore Journal of Tropical Geography1(1): 23-36, URA (2011). Conservation Guidelines for Historic Districts. S. Urban Redevelopment Authority.
③ "1989 marked a milestone in Singapore, when 10 conservation areas in the historic districts of Chinatown (Telok Ayer, Kreta Ayer, Tanjong Pagar and Bukit Pasoh), Little India, Kampong Glam, Singapore River (Boat Quay and Clarke Quay), Cairnhill and Emerald Hill, with a total of over 3200 buildings, were gazetted for conservation". Source: http://www.ura.gov.sg/uol/conservation/vision-and-principles/brief-history.aspx.

four main groups of conservation areas have been identified, including four historic districts, three residential historic districts, and eleven secondary settlements. In addition, four Bungalow areas and the Mountbatten Road Conservation Area have also been designated (URA 2011). Under the "3R" conservation principle——*"maximum retention, sensitive restoration, and careful repair"* (URA 2011), various conservation guidelines and adaptation measures have been applied based on different conservation degrees, historic significance and surrounding contexts(URA 2011).

- Top-down ethnic-based approach with strict conservation guidelines and facilitating regulations in core historic areas

In 1986 the Urban Redevelopment Authority released The Conservation Master Plan in which, Chinatown, Kampong Glam and Little India were designated as three historic districts. These areas are rich in unique ethnic cultures and activities. According to the Conservation Master Plan, one of the main objectives is to *"retain and enhance ethnic-based activities while consolidating the area with new and compatible industries"*(Henderson 2000).

Especially in the core areas rich in ethnic features, the top-down, strict conservation guidelines are applied to control the change of forms, functions, and land uses, etc., with "development charge" applied as a regulation to maintain the previous land uses. In addition, car parking waivers and strata subdivision are adopted to facilitate the strict implementation of conservation plan and for the maintenance of historic buildings by private owners(URA 2011). According to the survey conducted in 1994 by Lee(1996), for six historic areas[①], they successfully retained many "tradition trades" and "ethnic-based activities" by controlling the functional replacements in the first floor of historic buildings. Although it is criticized that in the case of Chinatown its implementation did not largely fulfill the conservation commitment due to the dominant tourism concern (Teo and Huang 1995, Chang and Yeoh 1999, Henderson 2003), Kampong Glam and Little India could be seen as successful cases in terms of preserving ethnic identity and ambience via strict conservation guidelines(Lee 1996, Chang 2000, Henderson 2003).

- Adaptive reuse of historic buildings with the intensification of land uses patterns from residential to mixed land use

As a city state, one of the most serious problems faced by Singapore is the scarcity of land,

---

① The six historic areas are: Telok Ayer, Bukit Pasoh, Kreta Ayer, Tanjong Pagar, Kampong Glam and Little India, and the survey was conducted in 1994Chang, T. and B. S. Yeoh (1999). "'New Asia–Singapore': communicating local cultures through global tourism." Geoforum30(2): 101-115.

which made it impossible to adopt the "twin-city" conservation model of other cities (Heng 1995). In other words the adaptive reuses of historic buildings had to be considered in order to intensify the land use patterns.

The Urban Redevelopment Authority developed categories of planning parameters and restoration guidelines regarding the adaptive reuses for the gazetted historic buildings. They are based on various degrees of historic significance, including the "building use" for the historic districts, the "rear extension" of residential historic districts, "service lane" and "restoration" versus "development" options for the secondary settlements, "new extensions" and "subdivision of land" in the category of Bungalows conservation(URA 2011). The thorough regulations successfully guided the adaptive reuse and new developments. By 1994, a field survey in six historic areas conducted by Lee(1996) has shown that *"most of the shophouses have turned from wholly or partially residential use to commercial use, with retailing on the first level and offices on the upper levels"*, which has also facilitated the requirements for tourism (Chang and Yeoh 1999) and intensification of land use in the central areas towards a higher commercial land use patterns from residential ones(Lee 1996) .

- Heritage revitalization through private developers facilitated with allowable maximum plot ratio of the Envelope Control Sites

To facilitate the new developments or extensions of the Envelope Control Sites[①], the "Sale of Sites Programme" was issued in 1989 (Heng and Chan 2000, Zhu, Sim et al. 2006), which embarked on new conservation approaches. Regarding the development of new projects or extension of the Envelope Control Sites, the development of Far East Square and China Square is a pilot project. The block of Far East square was designated as conservation area in 1997. After land acquisition and selling to developers for commercial uses, rather than from the funding of the government, 61 listed shophouses were physically conserved by developers and renovated for high end catering businesses, together with Fu Tak Chi Temple which was restored by private developers and renovated into a museum (Zhu, Sim et al. 2006). As plot ratio compensation, the adjacent new commercial development project——China Square was approved by URA under a *"maximum permissible"* plot ratio(URA 2011).

Other examples could be found in the Conservation Plan of Boat Quay and Clarke Quay, etc. (Savage, Huang et al. 2004, Yip and URA 2008). Nevertheless, there still are wide discus-

---

① According to the Conservation Guidelines for Historic Districts. URA (2011). Conservation Guidelines for Historic Districts. S. Urban Redevelopment Authority. Envelope control sites means *"vacant lands and buildings located within Conservation Areas, but not designated for conservation"*.

sions on the urban conservation policy and strategy of Singapore. In terms of the authenticity, scholars question the tourism-dominant, "Disney Style" or "Theme Park" heritage conservation modes, which are causing the vanishing of original street lives and activities(Kong and Yeoh 1994, Teo and Huang 1995, Heng and Chan 2000, Heng and Quah 2000). In terms of conservation policy decision making under the dilemma of conservation versus redevelopment and locals versus tourists, it is criticized that priorities have been primarily given to the top-down redevelopment and to the tourism image making while lacking concern for the public and local communities(Kong and Yeoh 1994, Yeoh and Huang 1996). Furthermore, from a property rights perspective under the institutional framework, Zhu, et al. (Zhu, Sim et al. 2006)argued that from 1960s until 2000 through a series of institutional changes on land property such as rent control, land acquisition, etc. the conservation and redevelopment of historic areas has been controlled by government, hence excluding stakeholders from public and private sectors from the decision making process of regenerating Niucheshui (Chinatown area).

2) *Approaches adopted for urban conservation in historic areas in Japan*
Japan has an extensive array of natural and cultural heritage and its legislation of historic buildings protection can be traced back to 1890s after the Meiji Restoration in 1868(Ketelaar 1990). After World War II, Japan started to restore and conserve urban heritage. To date, after several decades of practice, various approaches of urban and town conservation were developed by Japanese governments and researchers.Some examples include community making—"*machi-zukkuri*", "town making" (Akagawa 2014), sense of identity—"*furusato*",the fund and trust system,and among which, "*Machi-zukkuri*" community empowerment approach is well-developed in Japan and could be adaptable to other Asian countries.

- Preservation association initiated community empowerment approach——*Machi-zukkuri* approach

In community empowerment approach, conservation is normally initiated by local communities or NGOs, and local residents are encouraged to participate in the conservation activities. With the financial and technical support from government or organizations, not only the physical environment is preserved and promoted, but also the identity of the community and the sense of place could be enhanced. This approach was proven to be quite effective for dilapidated old towns and residential communities with the government giving priorities to economic growth.

In the years after the war rapid development, citizen movement rose, and the Japanese government was criticized for giving priorities to economic growth over "citizen's health and lifestyle"(Siegenthaler 2003). The case of Tsumago post town, as *"Japan's first citizens' movement to promote local historic preservation"*(Siegenthaler 2003) might provide a deep insight into this approach. Tsumago is located in Nagiso, Kiso District, Nagano Prefecture, Japan. In the 1960s, local communities of Tsumago realized the problem of population loss, especially of the younger generations. Therefore, the Tsumago Protectors' Association was established, under whose efforts, a campaign to preserve the town and enhance its cultural character was launched in 1968 (Siegenthaler 2003). Cooperation among local governmental authorities, experts of preservation, architecture and tourism, as well as local residents was motivated by the Tsumago Protectors' Association. Committees were set up to manage the development and to conserve the traditional urban fabric as well as identities, and nearly all the residents even agreed that they would not sell, rent or demolish their houses. By employing the term "*furusato*", which brings the nostalgic feeling of "native place" or "hometown" (Akagawa 2014) and the image of Japan's "pre-modern past", "rurality", "craftsmanship", and "tradition" (Siegenthaler 2003), rural tourism industry has prospered in Tsumago since 1970s. A local common building "*kôminkan*"—— "citizen hall" was established, which facilitates the widespread adult education in Japan, accommodates the roots of the preservation campaign in Tsumago, and fosters the community involvement(Siegenthaler 2003). These successful actions resulted from the combined efforts of the Tsumago Protectors' Association, local residents, government, institutions, and the "*kôminkan program*" have not only enhanced community vitality, cultural identity, sense of place and authenticity, tourism economy, etc.; but also successfully attracted people to return(Siegenthaler 2003, Akagawa 2014). Community empowerment has become one part of the citizen movements since 1960s, in 1975 Japanese government adjusted the previous 1950 protection laws for cultural capitals (Henrichsen 1998, Siegenthaler 2003, Issarathumnoon 2004). As a result, Tsumago post town was listed in one of the first groups of new preservation districts for "Important Preservation District for Groups of Historical Buildings"(Siegenthaler 2003).

2.4 Urban conservation in China

2.4.1 The history of urban conservation in China

Compared to most countries in Europe, the study of urban conservation began relatively late in China. The actual study with modern significance could be traced back in 1922 when the first academic institution for protection of cultural relics in China was established, namely the Institute of Archaeology in Peking University. In 1929, Society for the Study of Chinese Architecture（中国营造学社）was founded by Zhu Qiqian（朱启钤）and later joined by Liang Sicheng（梁思成）and Liu Dunzhen（刘敦桢）, who used scientific and systemic method to carry out the conservation of ancient Chinese architecture (Wang, Ruan et al. 1999). In 1930, the Chinese Government publicized the《古物保护法》, in which there are some regulations for urban conservation. In 1948,《全国重要文物建筑简目》was published, which was edited by Liang

Sicheng, who was working in Tsinghua University at the time. According to Chen, this achievement is the foundation of the first list of nationwide Cultural Relics Protection Units(Chiu and Moss 2007). After 1949, the development of urban conservation could be divided into three stages.

1) *First stage: National Cultural Relics Protection Units*
In the first stage which is from 1950s to 1970s, the urban conservation in China was only focused on individual ancient buildings and cultural relics. In 1961, the Provisional Regulations on the Protection and Control of Cultural Relics《文物保护管理条例》was announced by the State Council. In the same year, totally 180 sites were listed as the first batch of National Cultural Relics Protection Units, including palaces, temples, tombs, monuments, and so on. Because of the limited understanding of urban conservation at that time, only the most significant ones In terms of historical and architecture value were listed, for example, the Forbidden city and Summer Palace in Beijing. The units on the list would be strictly preserved by law and only can be used as museums and pavilions. Since then, this policy became one of the most important approaches to conduct urban conservation in China. Until now, after the Central Government announced the seventh patch of National Cultural Relics Protection Units in 2014, there are totally 4,296 units nationwide. Meanwhile, at the provincial, municipal, and county level, local governments may also list the most significant historical buildings and monuments as Cultural Relics Protection Units.

Due to the Cultural Revolution, only a few significant traditional buildings were well conserved in the first three decades after the people's Republic of China was founded and the urban conservation studies were difficult to be carried out during this period (Xie 2013).

2) *The second stage: Historic Cities*
The second stage began in 1980s after reform and opening-up policy. The defining events is that the central government listed 24 cities as "Historic Cities", also known as "Famous Cities of Historical Culture Value" in February, 1982. Advocated by Hou Renzhi(侯仁之), Zheng Xiaoxie(郑孝燮), and Shan Shiyuan (单士元), this conservation mechanism was proposed in 1982, with the intention to expend the contents of urban conservation. According to "Law of the People's Republic of China on Protection of Cultural Relics", " Cities with an unusual wealth of cultural relics of important historical value or high revolutionary significance shall be verified and announced by the State Council as famous cities of historical culture value." (NPCSC 2002). Up to now, totally 125 cities are in the list of national famous cities of historical culture value. It is also required that local governments at or above county level are responsible for the conservation planning for Historic Cities, which should be included in local master plan (Ibid.).

3) *The third stage: Historic Conservation Area*

In the third stage, urban conservation in China emphasized on urban historic area, streets, quarters, and neighbourhoods. Although the concept of historic conservation area was first introduced in China in 1982 when State Council announced the first list of National Historic Cities and officially mentioned in 1986 when the second list of National Historic Cities was released, the clear definition and specific conservation means were proposed in the 1990s. In June, 1996, the International Symposium for Conservation of Historic Area was hold in Tunxi District, Huangshan City, Anhui Province by China's Ministry of Housing and Urban-Rural Development (MOHURD, formerly known as Ministry of Construction), the Architectural Society of China, and the Urban Planning Society of China, which could be seen as the initiation of the third stage. In this symposium, the Old Street in Tunxi District was discussed as an example in order to explore a new dimension of urban conservation in China. The then minster Ye Rutang (叶如棠) of the original Ministry of Construction addressed that there were only a few cities could meet the requirement to apply for Historic Cities and got the supports for their urban conservation, but there are quite a number of streets, neighbourhoods, and villages are worth to be well conserved due to their important historical and architectural value (Fu 1996). This symposium drew a conclusion that a new approach for urban conservation was necessary for the purpose to protect the historic quarters so that they can embody the cities' history (Ibid.). According to "Law of the People's Republic of China on Protection of Cultural Relics", historic conservation area refers to "Towns, neighbourhoods or villages with an unusual wealth of cultural relics of important historical value or high revolutionary memorial significance" which "be verified and announced by the people's governments of provinces, autonomous regions, or municipalities directly under the central governments" (NPCSC 2002). As proposed in this conference, different with those strictly protected Cultural Relics Protection Units, buildings in historic streets, neighbourhood, quarters and villages should be allow to be endowed with new functions as long as their façades were preserved; meanwhile the conservation policies should also focus on the improvement of living conditions, infrastructures, environmental qualities (Wang and Wang 1998). In 1996, suggested by Qian Weichang (钱伟长), the Central Government set up a special fund for the conservation and maintenance of historic areas; 16 historic areas were funded in 1997 (Ruan and Sun 2001). According to Ruan, the conservation mechanism of historic areas added a new dimension and completed the system of urban conservation in China (Ruan and Sun 2001).

2.4.2 Current conservation strategies of urban historic quarters in China

In China, studies on the conservation of urban historic areas can be categorized in the areas of conservation theory(Ruan and Lin 2003),conservation policy and planning principle(Zhang 1999, Ruan and Sun 2001, Su 2010), urban renewal and urban design approach(Wu 1991, Lu

1997, Shin 2010), adaptive reuse of historic buildings(Chang 2003, Ruan and Zhang 2004, Wang 2009).The conservation plan of historic quarters or streets in China's cities is under Urban Regulatory Plan provided by Municipal government. Local government is highly involved in the policy making and implementation of conservation plan. Generally the following approaches present as the mainstream conservation and development patterns adopted by local governments.

1) *" Spot-line-area" conservation control approach*
Luo Zhewen (罗哲文) (2003) proposed the idea of "conserving the historical cultures in different ways based on their different scales and circumstances". The three-leveled conservation approach emerged, which is usually termed as "spot-line-area" approach. In this approach, the important historical buildings (spots), such as National or Provincial Protected Culture Relic Units are strictly conserved as museums or pavilions. Meanwhile, the façades of main streets (lines) of historical areas are also restored, with the street converted to commercial streets in most cases. The surrounding areas are controlled by planning as buffering zone and new development is restrained in these areas. This approach is often adopted by cities or towns to assure the important historical buildings and their surrounding environment; and to attract economic development of surrounding areas, such as those commercial areas around famous historic site. In most cases, functional restructuring is usually accompanied with this conservation approach. After the conservation, the whole area could be designated as tourists' destinations (for example: Lijiang historic town). Many researchers have criticized that this approach conserves historical area as theme park and the original identity and everyday life are threatened by the tourism activity and commoditization process(Xu and Tao 2001, Bao, Qin et al. 2005, Al-hagla 2010).

2) *Listed building conservation with redevelopment of surrounding area approach*
The second approach is driven by the economic growth in real estate. On the one hand, the important historical buildings are restored, which is similar with the first approach, while their function could be changed. On the other hand, the less important buildings or buildings with low architecture value are usually demolished and replaced by new developments. This approach conserves the specific parts of historical districts such as listed building and converts them to high class restaurant, café, hotels and galleries, etc., The demolished area is normally traditional residential districts, which suffer a lot in terms of community loss, home loss and even livelihood loss. This approach is criticized to pursue high economic profit while undervalue the social aspects and community life and cause social problems. "Xintiandi" project in Shanghai and "San Fang Qi Xiang" project in Fuzhou might provide examples for this approach (He and Wu 2005, Wai 2006, Yao and Zhao 2009).

3) Adaptive reuse approach

Accompanied with the rapid process of urbanization, cities undergo a large scale industrial relocation. In result, factories, warehouse and other industrial structures were left vacant in some old districts. Due to good construction qualities, some of these structures were not demolished but adapted to new functions to revitalize the urban area. Because reuse could reduce the development investment, revitalize the economy, and enhance the urban memories of past, this approach is popular in big cities with large amount of industrial heritage. Usually cultural industry and creative industry initiate the reuse of industrial factories based on low rent. The adaptive reuse can be seen in forms of artists' studio, exhibition hall, and design workshops, etc. The entire surrounding can be latterly revitalized, which somehow drive artists to flee away due to the increased rent. "Red Town Project", "the Bridge 8" in Shanghai and "798 Project" in Beijing could partly demonstrate the success of this approach(Tan 2006, Zhou and Shen 2008, Wang 2009). Especially in Shanghai, many old industrial buildings have been successfully reused to accommodate creative industry. Effective organizations and flexible principles regarding adaptive reuse have been released by municipal government. For example, the "Three Unchanging Principles" was proposed by Shanghai Creative Industry Center to promote creative industry, which declared that ownership right, the major building structure, and land use should be kept as unchanged(Wang 2009).

4) Bottom-up approach

As a centralized country, urban conservation practices with a top-down approach are often adapted by local governments. However, in recent years, there emerged a new trend which is to carry out urban conservation via a bottom up direction. This bottom-up approach has its own characteristic. In several cases, local neighborhoods are empowered in the process of participation in urban conservation practice. Local residents' basic needs and requirement will be collected to be considered in the conservation plan. However in many cases, local residents choose to participate only if the top-down approach threatens their livelihood. Therefore, it might be termed as passive bottom-up approach, for example the conservation practice in Nan Luoguxiang. However, there emerges increasing cases showing residents' initiative to participate; for example, residents show a strong willingness to community participation under the "Rent Gap Seeking Regime" in Tianzi Fang. Recent years, this approach attracts more and more researchers' attentions after several successful cases. For example, conservation projects of Tianzifang and Taipingqiao neighborhoods in Shanghai and the conservation program of Drum Tower Muslim Quarter in Xi'an can all be put into this category (Guo-zhao 2007, Yang and Chang 2007, Hai 2008, Zhang, Wang et al. 2008, Hai 2009, Shinohara 2009, Wang, Yao et al. 2009, Shin 2010).

5) Compensation for demolition in conservation plan

In general, there are three major types of compensation for house demolished based on local urban conservation plans: monetary compensation; house exchange based on property value, and house exchange based on flat area. Monetary compensation is a one-time deal, and locals usually fail to afford a house within surrounding area. House exchange methods usually relocate residents to fringe area, where there lacks facilities and employment. Recently, property-right exchange has been adopted in Si-nan Road historic building conservation, which provides chances for people to stay in their neighborhood nearby.

2.5 A summary of current research approaches on urban conservation

With the ever expanding scope of urban conservation and the burgeoning of cooperation between different academic disciplines such as architecture, urban design, urban planning, geography, and social science, studies on urban conservation are contributing to the interdisciplinary understanding of this research topic. Current studies of urban conservation present fivedimensions, which can be seen as follows:

2.5.1 Approaches in the cultural and historical dimensions for historical sites

After the Second World War, the scale and type of conservation objects grew from single historic monuments to groups of buildings, markets, towns, etc. (Larkham 1996, Tiesdell, Oc et al. 1996, Hobson 2004, Rodwell 2007). Unlike architectural conservation, urban conservation is more complicated due to complex contexts and the larger scale involved. Ashworth (1990) points out that during the change from architectural preservation to the conservation of historic areas, the concerns have shifted from simple physical architectural preservation to comprehensive urban enhancement. For instance, the considerations of the function of these areas and the creative use of historic buildings have been introduced into the previously simple physical preservation-based measures. More importantly, the rises of the characters of conservation areas inspired the adaptation of several research topics on the value of the conservation substances as follows.

1) *Townscape*
In Patrick Geddes' (1949) opinion, each city is uniquely evolved with diverse social and cultural contexts, with which contemporary planning and architecture should stay in harmony. This idea was shared by Gustavo Giovannoni's "coexistent city" which is featured by the "mutually supportive and harmonious coexistence" of both modern and historic features(quoted from Rodwell 2007), and later by Gordon Cullen's(1961) "coherent townscape". Via analysis of serial vision, place and content of the built environment, Gullen(1961) suggests that "the fabric of towns: color, texture, scale, character, personality and uniqueness" could be identified and compared with each other. Since the notion of townscape was presented, researchers (Conzen 1973, Slater 1984, Whitehand 1990) have employed the concept to carry out urban conservation studies thus

becoming an important instrument to manage and control urban conservation and renewal.

2) Urban morphology study

According to Sauer's (1925) book The *Morphology of Landscape*, the term of morphology first appeared in the biologic science in eighteenth century. It was adapted to study the organic forms and their structures.Urban morphology study was promoted by Conzen (1960) in *Alnwick Northumberland: A Study in Town Plan Analysis.* Morphology extends the traditional philosophical thoughts with the indentifying and emphasizing the importance of structural elements and their development sequences.According to Gu (2001), the dominate aim of morphological analysis is to recapture and maintain "the integrity of the urban texture". Based on this understanding, morphological analysis could be wildly adapted in urban study to reveal and maintain the structure and hidden order of urban form.

With the growing research interest of conserving the character of urban areas, much attention has focused onthe decision-making regarding the conservation of urban areas.Evidently, the decision is based on the significance of urban areas and the visibility of urban landscape. And urban morphology could provide the effective assessment tool in terms of the following aspects such as the identification of historic development layers, explanation of the importance of specific elements and selection of conservation area (Zheng 2006, Whitehand , Gu et al. 2011).

3) Typological approaches

Typological studies and contextual studies are developed by architects and urban researchers, which provide distinctive perspectives for understanding the urban form (Gu and Song 2001, Gu 2001). Caniggia's (1979) research, for example, examines the evolution of the urban fabric by identifying the original, traditional structural forms, which he termed as "the first building" types. In this same study, Caniggia also established a series of principles explaining how "first building" types have been changed and adapted. Caniggia argues that there is an optimum building type which can be determined by current social, economic and environmental conditions.

As mentioned earlier,each city is unique in terms of social, culture and historical context(Geddes, LeGates et al. 1949). However, based on the idea of urban typology, conservation areas and their components can be further sorted into more specific categories. Once the relevant typologies have been indentified, it is possible to develop and define special tools and approaches for adoption by urban conservation practice. In this way, urban conservation methods could be fitted into more general planning approaches for different typologies as well as help identify effective approaches for urban conservation practices. In other words, urban typology studies might bring urban conservation closer in alignment with urban planning strategies thereby serving as a facilitator between urban design, urban planning and urban conservation.

4) *Environment behavioral studies*

Studies on the link between human activities and the physical form focus on how people perceive the physical environment and how they behave in it (Rapoport and Silverberg 1973, Whyte and York 1980, Rapoport 1990, Lynch 1992, Gehl 2011). For example, in Lynch's series of studies, he uses mental maps as a tool to represent an individual's perspective of a place and he suggests using "edge", "node", "path", "landmark" and "district" as five key elements of space to conduct the urban form analysis. He develops the notion of imageability of a place, which means "the ease with which its parts can be recognized and can be organized into a coherent pattern" (Lynch 1992). The study of environment behavioral and imageability also provides a means to analyze the contents of urban conservation.

In summary, in the cultural and historical dimensions for historical sites efforts are taken to explore the cultural, historic and aesthetic value of conservation substances. Approaches in this dimension are mainly focusing on assessing the tangible values of physical fabrics, such as the hierarchy, structure and patterns of the physical forms, albeit the assessment tools are to a certain extent relied on the available historical and archaeological materials and findings. Thus approaches in this dimension are weak in exploring the intangible aspects of heritage value such as sense of place and everyday life of local residents.

2.5.2 Approaches in economic dimension

1) *Obsolescence and revitalization in historic quarters*

Obsolescence is a term revealing the extent of uselessness (Lichfield 1988, Larkham 1996) or diminished utility (Tiesdell, Oc et al. 1996). The analysis of obsolescence indicates some potential reasons for the reduction of utility. There are five types of obsolescence: structural obsolescence, functional obsolescence, economic obsolescence, rental obsolescence and community obsolescence, among which structural obsolescence, functional obsolescence and economic obsolescence are the most important factors to the built environment (Lewis and Weber 1965, Sim 1982). Litchfield (1988) identified two other factors, that is, "locational change" and "environmental unsuitability". Tiesdell (1996) later added "image obsolescence", "'legal' and 'official'" obsolescence and "relative or economic obsolescence", with the second one addressing the blight by official zoning and designation, and the latter addressing the competition for opportunities of investment between the fabric of historic buildings and other areas.

The analysis of obsolescence provides a perspective to understand the reasons of the decay of urban historic areas. Except for physical and some functional reasons, it is clear that government policy and action could influence the rate of obsolescence (Lichfield 1988). The obsolescence of a historic area can also be seen in a filtering down process: obsolete areas will attract those of low socio-economic status.

Based on the analysis of obsolescence, Litchfield (1988) discussed the life cycles of urban fabrics and argues that compared with obsolescence and renewal, conservation will improve the "physical", "functional", "locational" and "environmental" obsolescence in the long run. Accordingly, how to remedy the obsolescence, in the same time to extend the useful life circle, and in the end to revitalize of the historic buildings became a major concern for conservation of historic areas(Lichfield 1988). At the same time, the cycle of de-investment and reinvestment also leads to the need of revitalization (Doratli 2005, Gottdiener and Budd 2005).

The revitalization of historic quarters contains three aspects: physical, economic and social revitalizations. The first two aspects and their relationship have been discussed widely. Physical revitalization and economic revitalization are interrelated and inter-constrained. Physical revitalization usually takes a short time but it will drive or constrain economic activities in the long run, and the maintenance of physical revitalization requires continual economic support. Similarly, short-sighted economic revitalization has been implemented frequently in pursuit of rapid economic growth, which is accused of destroying the traditional urban fabric. Historic buildings and areas in the long term need investment from property occupation for maintenance and renovation. Tiesdell (1996)addressed the importance of economic revitalization over physical and social revitalization: "in the longer term, economic revitalization is required because ultimately it is the productive utilization of the private realm which pays for the maintenance of the public realm…in the absence of large-scale public subsidies, historic urban quarters need to establish and maintain their positions as centers of production and /or consumption." Accordingly major efforts have been implemented towards the improvement of economic revitalization in order to generate economic growth and increase the demand for use of historic areas.

Economic revitalization will ensure the maintenance of physical revitalization; an attractive physical fabric and environment, in turn, will encourage public activity and enhance "the sense of place". However, the unique characters of history, culture and tradition of a historic area will also attract the economic use and investment (Elsorady 2011).

2) *Economic restructuring: influences and opportunities*
During the last several decades, there emerged great economic change all over the world. With economic restructuring and industrial relocation on a worldwide scale, many centers of produc-

tion have transformed into centers of consumption (Castells 1985, Harvey 2003) and many historic quarters have lost their original functions (Lichfield 1988, Larkham 1996, Tiesdell, Oc et al. 1996). Compared with other urban areas which have also undergone economic change, this global economic change especially influenced the historic areas. Firstly, under such economic change, many traditional historic quarters are no longer "*self-sufficient*" functional sectors and the traditional functions of historic quarters are prone to abandonment due to the difficulty of adapting the old physical forms and infrastructures to new technical requirements. Secondly, it is more difficult to deal with the economic change under restrictions of physical preservation (Tiesdell, Oc et al. 1996). Accordingly, with economic restructuring, many historic quarters are faced with decay. For example, during the 1970s, de-industrialization in many western countries caused massive unemployment and decay of industrial cities. In this process, many historic quarters in industrial cities were confronted with economic obsolescence (Lichfield 1988, Larkham 1996).

Meanwhile, Tiesdell (1996) argues that the economy of the historic quarter belongs to the economic dynamism of the city and insists that the historic urban quarter should be included within the context of the city. This means the economic revitalization in the historic quarter is an exigency to be addressed in urban development under the context of economic restructuring. The international economy and processes of globalization increase transnational cooperation and competition between cities (Castells and Hall 1994). In many historic cities, historic quarters are "repositioning themselves into the global economy" and searching for new functions and activities to revive their economies (Haughton and Hunter 1994).

3) *The "demolition vs. reuse" agenda and functional initiatives for reuse*
Before the introduction of area-based conservation, the dominant preservation perspective on historic buildings prohibited change and is accused of being responsible for incurring maintenance expenses due to the fact that strict limitation of the use of buildings restrained the income potential of owners. (Ashworth 1991, Larkham 1996, Hobson 2004). Area-based conservation, on the other hand, provided a new view which aimed to reinvent reasonable use and function to the buildings, revitalize the economy and enhance the historic area. In this sense, as compared to preservation, conservation could be argued as an enabler of reuse and economic development.

According to Tiesdell (1996), the economic growth of historic areas can be gained through three basic methods. The first one is functional restructuring which replaces the existing activities and functions with new uses and activities. The second one is functional regeneration which maintains and enhances the existing functions and improves the competitiveness the area's existing functions and uses. The third one is functional diversification which brings in some new uses to the historic urban area and can complement existing ones.

New activities are introduced to historic urban revitalization in order to address increasing obsolescence. Tourism-led, housing-led and culture-led revitalization are three main forms of functional initiatives, which aim to promote the area's economic growth and increase the use of its historic legacy thereby reducing image obsolescence. It usually uses the "functional diversification" or "functional restructuring" methods to change the area's economic base. As a result, new functions may totally "replace or complement" the traditional local use and activities (Tiesdell, Oc et al. 1996). There is an abundance of studies on the functional reuse of historic areas ranging from inner city, industrial factories, market towns, waterside areas (Tiesdell, Oc et al. 1996, Pickard 2001, Hobson 2004, Madgin 2009, Zhang and Chen 2010).

As can be seen from many practices, the functional restructuring aim is to increase the land value and rental value of properties, as well as to attract higher value use (Tiesdell, Oc et al. 1996). Functional restructuring can change the area's economic base and function and at the same time replace the activities of the historic urban area and its users. This process is known as gentrification (see approach "gentrification").

In summary, economic restructuring analysis provides an approach to explain the decay of the historic quarter and introduces opportunities and directions for the further development of historic areas. The approaches which examine conservation topics on "obsolescence and revitalization", "demolition vs. reuse agenda" and "functional initiatives" can be characterized as adaptation of radical economic analysis, such as "supply - demand relationship " and "life cycles of building" to explain the phenomena on urban conservation and to propose positive urban conservation solutions. However, these approaches in general lack the concerns of cultural, social and public aspects which have led conservation into an unbalanced situation: economic revitalization becomes a prime concern in decision-making and policy-making.

2.5.3 Urban management dimension: conservation planning and management approaches

As urban conservation is increasingly influenced by other urban categories such as land use, housing, real estate, transportation, and so on, scholars argue that conservation should be an integral part of urban planning (Ashworth and Tunbridge 1990, Tiesdell, Oc et al. 1996, Cohen 1999, Hobson 2004). Washington Charter (ICOMOS 1987) declares that "the conservation of historic towns and other historic urban areas should be an integral part of coherent policies of economic and social development and of urban and regional planning at every level." Other discussions can be seen on the extent that the legislation planning is effective to conservation, including such issues as the unlisted heritage buildings, the limitations of planning, and the professional distinction between conservation and planning (Punter 1987, Hobson 2004).

1) *Conservation control and regulation*

Conservation control is very important to both architectural conservation and area-based conservation. It prevents potential damage to historic buildings and historic areas in the process of urban renewal and development. Basically three methods are used to implement conservation control: listing, designation and larger scale planning control (including buffering zone and construction control area). According to Hobson (2004), conservation control has some limitations: (1) the meaning of "value" has been extended wider than "special architectural or historic interest", while the control guideline usually shows the "dominant value standard" and political interest; (2) rather than single building preservation, the "wider spatial relationship" and "the character of an area" are difficult to indicate in management guidelines; (3) listing has strong regulatory power while designation control is weak due to the variable complexities of an area's context. However, in his study on conservation controls in Australian cities, Nankervis (1988) argues that conservation in the case of the Australian state of Victoria has adapted a sensitive control which pays enough attention to protect private property values——it has been widely accepted and produced positive effects in control management.

2) *Conservation policy, politics support and public intervention*

Tiesdell et. al. (1996) suggests that policy-initiated conservation planning has emerged which concerns the revitalization of historic areas especially from the perspective of economic development. As historic areas are closely influenced by factors such as land use, housing, transportation and economic activities, a policy framework based on the cooperation between different agencies from various local and national scales is necessary in conservation management. Recently there have been greater calls towards sustainable development through a long-term policy in the conservation and regeneration of historic towns and areas (Strange 1997, Madgin 2009). Current studies related to long-term conservation planning can be seen to include the following aspects: analysis of pressure and options/SWOT analysis for policy making (Strange 1997, Doratli, Hoskara et al. 2004); evaluation and assessment for further improvement of policy (Peel 2003, Kocabas 2006, Lee, Lim et al. 2008); and the explanation of importance of conservation policy based on specific cases (Ashworth 1984, Lee 1996, Madgin 2009).

Politics in urban conservation is closely related with property rights, ownership, and funding. As Madgin (2009) points out with two case studies that "funding financial support and securing an owner whose aims were concurrent with local and national government policy was vital to secure the restoration and re-use of the historic buildings", in urban conservation, political support is especially important to balance the interests between different groups, among which the conflicts between economic profit, public good and private right are always the focus of discussions.

Given the protection of private property rights, public intervention is used to "balance collective welfare against private interest and by including consideration of certain negative externalities into the price mechanism"(Tiesdell, Oc et al. 1996),especially when demolition of historic buildings based on pursuing private profit is discussed. Free market economy pursues the highest land use value based on the relationship of supply and demand, which usually advocates demolition rather than the reuse of historic buildings. This dominant thinking lacks a social concern and leads to the unawareness of social needs. The destruction of historic urban areas usually results in the loss of public welfare (Tiesdell, Oc et al. 1996). Public intervention into private free market is necessary to protect the public good, which can lead to the reasonable reuse of buildings rather than demolition.

3) *Place management: place marketing and place making*
The concept of place has been introduced to the field of conservation and the ideas of "place making" and "place management" have become important concerns in urban conservation: "Place making is now at the heart of conservation" (Worthington, Warren et al. 1998).A sense of place has become a significant evaluation factor in conservation management (Hobson 2004). Area enhancement and a sense of place have become emerging challenges to conservation (Larkham 1996).

Place management of urban conservation areas usually has two processes: place marketing and place making. Place making is a process to "reconstruct the image of a place for both visitors and local people and to encourage the local residents and business community to achieve regeneration" (Tiesdell, Oc et al. 1996). Through the enhancement of the quarter's cultural atmosphere, unique architectural character and associated activities, the entire area can attract and support both new uses and local people.

Nevertheless, most of the schemes simply focus on how to attract new uses such as tourism and usually conduct a place marketing process (Kearns and Philo 1993) and uses the "functional diversification" or "functional restructuring" (Tiesdell, Oc et al. 1996) methods to change the area's economic base. Accordingly, new economic activities such as tourism may "replace or complement"(Tiesdell, Oc et al. 1996) the original uses of buildings and economic activities.

As physical and economic revitalization have been paid overwhelming attention, a failure to address the issues of culture, social values and community need has largely resulted in an imbalance of the social environment such as loss of identity, collective memory and traditional activities, etc. Since both urban functions and people's social activities contribute greatly to the identity of the historic area which is frequently overlooked by economic value through place marketing process, place marketing is frequently criticized as homogenizing urban images and destroying the "sense of place".

Harvey (1989) criticizes that the uniformity of city images had largely reduced their distinctiveness and argues that a similar experience which a visitor can experience in another place is less attractive. However, an historic quarter with a vibrant community could interpret itself well, abounding in everyday life which generates distinction.

A perspective of everyday life (Lefebvre 2005, Liu 2008)could explain the distinction of sense of place and identity in historic areas. Lefebvre (2005) identifies everyday life as a repetitive and cyclical time process and tradition and also a linear time process of ever changing and developing and modernization. The complexity of everyday life lies in that no matter whether in the historical, philosophical, economic or social process, it could not be understood as a single linear process of accumulation and development, which means everyday life embraces tradition. Everyday life is the intersection of different aspects and bears the richness of every part. In this case, it is difficult to categorize everyday life into any specific way of production and everyday life belongs to collective culture (Liu 2008).

Another related question is the authenticity of place (Zukin 2009). Zukin argues about the authenticity of place based on two meanings: absolute authenticity of existence and creative renovation, the second of which is now being increasing realized. With the increasing recognition of the of "living character" and "permanent character" of cultural heritage, "the spirit of place" is proposed in 2008 ICOMOS as a constantly "reconstructedprocess" which change over time and in conjunctions with specific contexts and involved people (ICOMOS 2008). In light of this, the spirit of place highlights a more collective value system which not only is embedded in the permanent heritage value, but also is embracing the living character, community's needs and a continuous creation of values along with the process(ICOMOS 2008, Zukin 2009, Wells 2010). Inaddition, a more inclusive participation of stakeholders whose collective actions in the conservation practices will contribute to a more collective spirit of place which can be understood and shared by diverse groups with different grounds (ICOMOS 2008) .

Studies on place management perspective can be seen in discussions on conflicts between continuity of sense of place and change (Galway and Mceldowney 2006, Wang 2008), place making and public engagement (Chase 2000), the relationship between sense of place, identity and social and cultural values (Liang 2008, Ardakani and Oloonabadi 2011). Studies show that compared with technical conservation control and regulation, place management requires a comprehensive understanding of the urban context and multi-disciplinary cooperation to deal with the relationship between development and conservation of culture, identity, tradition, memory and community needs, etc. Accordingly, sociology, development control, public participation and partnership engagement are also necessary in the enhancement of place.

4) *Public participation in the process of decision making*
Public participation is an important mechanism in urban governance. It can help solve conflicts (Sanoff 2000) and ease the implementation (Creighton 2005). The approach of public participation provides new perspectives in the area of urban conservation. Studies have been focused on the role, opportunities and challenges of public participation in urban conservation (Sharif Shams 2006, Ping 2007, Zhang 2008). Firstly, in terms of participants, the scale extends largely, as Atlee (2008)argues that "Those who are affected by a decision have a right to be involved in the decision-making process". Creighton (2005)proposes a "multiple techniques" approach, which includes policy committee, technical committee, NGOs, citizen advisory committee and media.

Secondly, in the urban conservation decision making process, public participation has played an increasingly important role. The process of public participation should be integrated into the process of decision making——identifying problems, negotiating alternatives, applying assessment, and providing remedy measures, etc. ——as such, a good understanding of history, identity, community context, needs, interests and values can be integrated into the process of decision making (Creighton 2005).

5) *Community development based conservation*
Community development aims to provide the community with empowerment to resolve the problem of (in)equity through a bottom-up community-oriented development process which usually includes: identification of community problems and issues, development of community organizations, creating visions and setting objectives, making development plans and regulations, implementations and evaluations (Wideman 1998, Bhatta 2009). The community development approach provides a perspective which examines the relationship between conservation and community development. Some scholars see urban conservation as a way to promote the sustainable development of community (Snyder 2008, Bhatta 2009, Elsorady 2011). In other words, by integrating community needs and resources with conservation and/or tourism plans,

the community will be provided with better opportunities for future development with less impact on their identity and daily life and, at the same time, issues of (in)equity can be addressed (Salazar 2006). Community participation and interpretation of the different actors involved in the goals of conservation programming are the keys in a community development plan (Salazar 2006).

The community development approach also reveals the impact of conservation planning on the life of the community (Elsorady 2011), which will in turn influence the historic area's future development. For example, many historical quarters are increasingly dependent on tourism for regeneration; hence, the day-to-day lives of the local people are continuously affected by tourism development, resulting in the emergence of severe problems and conflicts. Without prompt mitigation, such historic areas will suffer from decay. William F. Theobald (1998) points out that, beyond the social, cultural, natural environmental quality and the experience of tourists in local places, the extent to which local residents accept tourism, and their perception and attitude to the impact of tourism development are also important affecting factors to the sustainable development of tourism.

As a bottom-up approach, community development addresses the issues of participation, equity and multiple development choices. However, as most local organizations exist at the community level, it usually lacks the connections to government, agencies and other institutions that can provide proper guidance for implementation. Accordingly, the efficiency of plan implementation is reduced. In addition, the evaluation of conservation's impact indicates that, rather than a one-time setting of objectives, community participation should be included into the whole process of development.

In summary, among the five approaches related with conservation management, conservation planning and control provides a technical, regulatory and relatively top-down approach for conservation implementation. It is easy to adopt, while it lacks the historic and contextual view provided by urban morphological analysis and townscape study (Barrett 1993, Larkham 1996, Mageean 1999, Pickard 2001, Hobson 2004). Approaches, such as policy-led conservation planning, public intervention and participation, and place management, are characterized as relatively "soft" approaches, which promote participation and understanding of sense of place, compromise the top-down approach, and balance the conflicts among different interest groups(Larkham 1996, Tiesdell, Oc et al. 1996, Strange 1997, Creighton 2005). However, without a comprehensive communication and associational framework, the simple adoption of these approaches to conservation will influence the efficiency of conservation in terms of decision making and implementation.

2.5.4 Social spatial dimension: gentrification, social spatial reconstruction and replacement

1) *Gentrification and social replacement*

Gentrification is defined as "the transformation of a working class or vacant area of the central city into middle-class residential or commercial use"(Lees, Slater et al. 2013). Gentrification is an urban phenomenon emerging in the late 1970s and 1980s when inner city quarters were under functional changes or social changes to revitalize an area from urban decay. Generally, gentrification is one result of the revitalization of a historic area. It starts with housing-led revitalization but has now occurred in other sectors (Tiesdell, Oc et al. 1996). Tiesdell et. al (1996) describe the mechanism of gentrification: "building owners and other landlords seek to increase or maximize their profits by trying to attract higher value uses and/or tenants able to pay higher rents".

However, accompanied with functional change, there is usually social replacement. When a historic quarter is suffering obsolescence, pioneers come in and improve the conditions, which attract higher-income classes to live and consume in the area. With the real-estate industry actively involved in the buying and renting circle, landlords increase the rent and the original working-class with their convenience shops are displaced by the middle-class and their consumption tastes and habits (Appleyard 1979).

Gottdiener and Budd (2005) see gentrification in the perspective of capital investment. When an area is decayed, the property owner usually will stop investing money for maintenance and upgrading, which changes the area to a lower-rent district. As a result of disinvestment, the depressed areas suffer from further deterioration and have much lower property values. As some areas have attractive features such as being a historic area, having a central location, and possessing interesting buildings, real estate will reinvest in these areas and attract affluent residents by providing higher-class accommodation and services. Accordingly, "it is the process of disinvestment and re-investment results in cycles of decline and gentrification that afflict the housing stock of the city"(Gottdiener and Budd 2005).Other city marketing events, such as sports and festivals, will also result in gentrification.

At the same time, the increase in rent and price of goods and services accompanied with neighborhood gentrification displaces working class and low-income people, leading to social and spatial replacement. Thus community concerns and social conflicts emerge. Displacement, however, has different conditions. If it is a top-down replacement of all residents and functions, the historic quarter's sense of place will be largely damaged (Tiesdell, Oc et al. 1996) since both urban functions and people's activities contribute greatly to the identity of the historic area and its sense of place and the livelihood of residents will be a big challenge to society. However, Tiesdell (1996) argues that if the original residents and business are property owners who can stay, displacement of some tenants will not be totally harmful as the values of property increased with the sense of place retained.

In his paper "The hidden costs of gentrification: displacement in central London", Atkinson (2000) discusses the reason and effects of "gentrification–induceddisplacement". According to his study on central London, gentrification has both positive and negative effects. Even though urban management and interventions strategies have reduced some of the adverse impact on vulnerable groups, the displacement of social groups and original communities has caused big social problems such as the weakened diversity and affordability of the original communities and their surrounding neighborhoods (Atkinson 2000).

2) *Conservation initiated Urban Regeneration*
In contrast to gentrification, urban regeneration aims to regenerate a derelict urban area in order to achieve physical and economic revitalization, as well as social well being (Roberts 2000). Urban regeneration has many initiatives such as housing, tourism and environment. As many historic areas are suffering decay, urban conservation is usually seen as one component of urban regeneration, with special emphasis on the management of urban heritage. Urban regeneration requires multi-agency partnerships (Evans and Jones 2008), while current discussions on regeneration still emphasize heavily on the economic dimension. Sustainable development provides a multi-disciplinary approach to the study of urban regeneration. Discussion has been developed in the following aspects: the policy and partnership in sustainable urban regeneration (Couch and Dennemann 2000, Evans and Jones 2008); economic regeneration, tourism and regeneration (Owen 1990); gentrification and regeneration (Güzey 2009); the conservation participation in urban regeneration (Tarn 1985, Shin 2010); and the "rent gap seeking regime" based on conservation-led urban regeneration (Yang and Chang 2007). Recent studies show increasing concerns on the social dimension such as local needs, exclusion in the urban regeneration process (Chen 2010, Shin 2010, Uysal 2011).

3) *Social spatial approach and spatial restructure*
The social replacement which occurs as a repercussion of gentrification in historic areas can be understood from the perspective of social spatial approach. As the Human ecology perspective of human's biological adjustment——"struggle for survival" failed to explain many urban issues such as the de-industrialization of the 1970s, a social spatial approach emerged in the 1990s with a perspective that "the form of settlement space is related to the mode of organiza-

tion of the economy"(Gottdiener and Hutchison 1994).Social spatial approach provides a perspective to explain the spatial arrangement in newly developed and modern urban areas, while the adaptation of social spatial approach to urban conservation areas helps to explain issues of urban decay, redevelopment, urban renewal and social restructuring.

Many governments' reaction to the urban decay and deterioration of inner cities are criticized as "planned shrinkage" (Gratz 1995). Due to shortage of government investment, many traditional communities relied on rental properties as income generators, hence old neighbors moved out and got less connected with each other. With the shortage of infrastructure maintenance and a breakdown in the sense of community, many old quarters are intentionally decayed and later demolished.

Modern urban development has been criticized in pursuing maximum economic gain at huge social and natural costs. After economic restructuring, financial capital came into real estate, which has become one of the driving forces of urban development. The "bulldozer"——mass demolition of traditional communities——and the popular "new" development approaches have physically and socially reconstructed the urban and social fabrics, thus diminishing social and economic diversity as well as the diversity of urban patterns.

As the capitalist market pursues maximum economic growth, real estate is used as a mechanism to increase land value through a process of urban renewal: dis-investment and re-investment generates "cycles of decline and gentrification" which "undercut the social values attributed to physical fabric" (Gottdiener and Budd 2005). Meanwhile, through redevelopment and urban renewal, urban areas are transformed as the composition of the local population and activities change over time. With the mass relocation of local residents, new social spaces as well as social-spatial segregation are produced within cities.

In summary, gentrification and social spatial approach provide a social economic perspective to understand the economic, social and political context of urban development. These approaches raise the issue of displacement and social restructuring, thus reflecting the limitation of urban conservation in that the economic concerns which currently dominate conservation planning undervalues the social diversity and cultural identity of local communities(Urry 1995). Moreover, these approaches strongly emphasize the function of private capital and the free market and overlook the function of urban policy, planning and management.

2.5.5 Sustainable development as an integrated approach

Sustainable development means "*development that meets the needs of the present without* compromising the ability of future generations to meet their own needs" (WCED 1987). The definition of sustainable development has been developed through the concept of the "three pillars" model, which lists the environmental, economic and social dimensions as domains of sustainability (Keiner 2005). The term "to sustain", which implies the ability to mitigate present problems and provide opportunities for continuity without depriving the rights of future generations, challenges our understanding and practice of "the legacy of past" from a sustainable urban conservation lens. The concept of sustainable development as an approach to conservation planning brings new perspectives and requires new methods.

Contemporary conservation approaches focus on sustainable development and tend to address economical and environmental issues such as energy consumption, environmental pollution, recycling the building materials and so on and so forth. Existing literature shows that previous research on urban conservation focused more on technical issues with the intention to maintain the fabric of existing buildings and reduce harmful impacts on environment. For example, Brimblecombe & Saiz-Jimenez (2004) discussed how to minimize damage on culture heritage during the process of conservation by dealing with atmospheric pollution. In addition, several research studies propose recycling of building materials and adaptive reuse of historical constructers. For instance, Wong and Rong (2001) presented their research on adaptive reuse of industrial buildings in the conservation process. Savage et al (2004) also introduce adaptive strategies in the conservation of the Singapore River. In summary, these studies mainly contribute to understanding how resources, both environmental and natural, can be used while decreasing the impact on the ecosystem during the process of urban conservation.

Increasingly, the role of urban heritage conservation in promoting economic growth has been acknowledged, which is mainly in association with tourism development. In this age of globalization, historical quarters attract tourists from around the world due to its cultural uniqueness and historical value, which can help to boost the economy in both local and national levels (Tweed and Sutherland 2007). Together with culture tourism development, the economic sustainability of urban conservation has become a major research topic. For example, Al-hagla (2010) highlights the importance of cultural assets of historical areas and emphasizes the sustainable development of urban historical areas based on "their potential as cultural tourisms sites". Similarly, Nasser (2003) also addresses the sustainability issues in his research and concludes that tourism can have positive attributions to economic development and social stability for urban historical areas; potential disadvantages, such as conflicts between locals and visitors might be mitigated through urban management strategies.

The social dimension of sustainability emphasizes the issue of quality of life for all citizens through improved "base levels of material income and by increasing social equity, such that all groups have fair access to education, livelihood and resources"(Tweed and Sutherland 2007). Related to urban conservation, the most important task is to revitalize the old heritage quarters, keep their cultural authenticity and promote social cohesions. In terms of social value, collective memory of heritage is addressed as an important social value(Urry 1995) and has been developed as a indicator to drive urban conservation towards sustainable development (Ardakani and Oloonabadi 2011). In terms of heritage value identification, there emerges the changing perspectives of understanding heritage as "fluid phenomenon" of cultural practice and social process, and of defining heritage value with an emphasis on the value status dominant by the "framework of values of the time and place", hence asserting that heritage of contemporary use should be a inclusive process rather than just a "static set of objects with fixed meanings" with diverse groups of people participate in (Pendlebury 2008). In terms of urban governance, efforts can be found on addressing the role of collaborative forms of governance to the inclusive preservation of historic towns(Nyseth and Sognnæs 2013)

However, compared to environmental and economic dimensions, a literature review of urban conservation shows a lack of publications regarding sustainable approaches to promote social sustainability by reconciling the social impacts such as social justice and social cohesion through a comprehensive urban conservation program.

Nowadays, with stronger motivations to enhance the economic development and improve the physical environment of urban areas, social sustainability has remained ignored or less considered by local governments, especially in developing countries. It has been verified in many cases that without a solid foundation for social network, social justice and social well-being, urban conservation would be less able to achieve sustainability and the economic and environmental dimensions will be weakened progressively (Tianwei and Di 2000, Shao, Zhang et al. 2004, Ruan and Yuan 2008, Yu-fang 2011). Therefore, future research on urban conservation should be encouraged to adopt the sustainable development perspective. Furthermore, win-win solutions and sustainable conservation methods focusing on the social dimension are significant topics that merit attention by researchers engaged in urban conservation studies.

Chapter 3 Theoretical Underpinning

3.1 Urban social sustainability

Yiftachel and Hedgcock (1993, p. 140) provide a definition of urban social sustainability, which refers to that "the city as a backdrop for lasting and meaningful social relations that meet the social needs of present and future generations."

Urban sustainability has been a focal point of urban studies research since the 1990s. However, as Yiftachel and Hedgcock (1993, p. 139) argues, "the debate on urban sustainability has mainly focused on environmental and economic factors, often neglecting the vital social aspect and an evaluation of the role of urban planning." Contributions that examine and advance the social component are extremely infrequent in the literature on urban sustainability studies, as compared to the environmental and economic dimensions. Fundamentally, a complex interrelationship of social, ecological and economic aspects is involved in the conception of sustainability (Taschereau, 2001). The most possible reason is that urban social sustainability is more intangible than environmental and economic ones which can be easily shown in the promotion of infrastructures and statistic numbers (Taschereau, 2001). In practice, Hancock (2009) asserts that the hard infrastructures (severs, roads, utilities and other physical structures) are usually the development elements to be considered firstly, while the soft infrastructures (social relationships, culture, community organizations, etc.) are assumed to be naturally attached to the hard infrastructures.

To date, the negative consequences caused by the neglected concerns on urban social sustainability can be observed worldwide, especially in developing countries. Society isolation, increasing crime rate, social inequity and other consequences have hindered the city development to a large extent, which has drawn increasing attention on the importance of social sustainability among scholars, policy makers and planners.The following sessions will discuss two key indicators of social sustainability——social justice and social cohesion.

3.1.1 Social justice

Cuthill (2010) argues that social justice is the "ethical imperative" to social sustainability. Though social justice has been an eternal goal that social sustainability pursues, it provides the insights for urban planning in terms of social sustainability.

Firstly, the reflections on social justice in terms of planning outcomes will help to understand the issue of urban sustainability. In his book "Social Justice and the City", Harvey (1973) analyzes that, during 1960s to 1970s, urban space had become the center of decay and argues that the urban geographical and spatial phenomena of urbanism is the result of profit driven capitalism-embodied planning activities: capital accumulation and political control. By applying a political economy approach, the study explains the phenomena of the polarization of society and examines the inequality in planning as a result of income redistribution and political decisions (Fainstein, 1997; Harvey, 1973).

One the one hand, the political economy approach accuses planning schemes of increasing inequality in that the "redistribution of real income" through political decisions into the urban system provides profits to a small group of people (the rich) while damaging the interests of a larger group of people (the poor) (Fainstein, 1997; Harvey, 1973). On one hand, the spatial and geographic circulation of surplus value leads to spatial and geographical aggregation, such as the concentration of wealth production in the CBD. Moreover, class relationhas resulted in the congregation of the working class and the formation of ghetto areas, which explains the urban phenomena of the polarization of society and the alienation of space.

Secondly, during the 1970s and 1980s, many countries like the US and the UK experienced an economic and political transition. The centrally planned economies in most of the western countries collapsed and were replaced by so called "market triumphalism". Under the market economy, it was assumed that the market could define justice and rationality automatically, and, guided by this economy, "there is no need for explicit theoretical, political and social argument over what is or is not socially rational just because it can be presumed that, provided the market functions properly, the outcome is nearly always just and rational"(Harvey, 1992, p. 597). However, the increasing phenomena such as unemployment, environmental problems, enlarged disparities, the increasing social stresses and so on reveals the weakness of a market economy (Harvey, 1992).

In addition, there emerged increasing requests from the public with concerns about the environment and appeals for coordination: "voice of state regulation, welfare state capitalism, of state management of industrial development, of state planning of environmental quality, land use, transportation systems and physical and social infrastructures, of state incomes and taxation policies which achieve a modicum of redistribution either in kind (via housing, health care, educational services and the like) or through income transfers, is being reasserted"(Harvey, 1992, p. 597). Just as government intervention is necessary to complement the "hidden hand" in market economy, urban planning also needs certain levels of government intervention based on the reflections of modernist planning outcomes. Meanwhile evaluation standards in the dimension of social sustainability have also been discussed in many government policies. For example, the British Columbia Round Table on Environment (1994) addressed social sustainability regarding sustainable development.

Thirdly, during the social economic transition in the 1990s, a postmodernist approach re-conceptualized social justice and inequity (Fainstein, 2005; Harvey, 1992; Sandercock, 1998; Young, 1990). On one hand, the perspective of justice has shifted from "redistributive justice"to "an institutional condition that enables participation andovercomes oppression and domination through the achievement of self-development andself-determination", as Young (1990, p. 241) argues, "social justice must consider not only distributive patterns, but also the processes and relationships that produce and reproduce those patterns".On the other hand,the postmodernist approach supported a socially inclusive process rather than solely focusing on "substance and material outcomes" (Cardoso & Breda-Vazquez, 2007, pp. 384,389). To quote Young (1990, p. 39): "justice should refer not only to distribution, but also to the institutional conditions necessary for the development and exercise of individual capacities and collective communication and cooperation".

Healey (2003) points out that the interactions and implementation process of development plan and policy planning have produced "distributive injustice", as most of the policy and norms are more effectively implemented in aspects of environmental concernsunder the condition of private investment, given investments from public sector are comparatively ineffective in many urban development projects, especially regarding urban regeneration of decayed urban quarters.[①] This concern of social justice from aspect of public sector questions the current approaches of both "neo-liberals" and "postmodern" regarding spatial planning, and call for new "dimensions of driving forces" that bring forward opportunities and initiate more "strategic and spatially-integrated approachesto territorial development"(Healey, 2003, p. 104). As such, Healey puts forward the concept of "collaborative planning", which highlights the following four key ideas: 1) "interactive process", 2) participatory governance activity under complex institutional background and various driving forces, 3) "planning and policy initiatives for quality of places", 4) "moral commitment to social justice , especially as realized in the fine grain of daily life experiencesin the context of culturally diverse values about local environments andways of life"(Healey, 2003, p. 105).As such, social justice shall be evaluated in terms of both material outcomes and policy making process.

[①] "So 'urban policy', which provided the investment for 'urban regeneration', proceeded much of the time with little attention to the policies and norms contained in development plans. The consequence was that 'place quality' was more neglected in areas dependent on public resources, which were typically poorer neighborhoods, than in areas dependent on private resources" (Healey, 2003, p. 104).

Prior to Healey (2003), Harvey (1992)and Sandercock et.al (1998)point out the importance of justice in an everyday form. Harvey (1992, p. 598) addresses the everyday meaning of justice, "to which people do attach importance and which to them appear unproblematic, gives the terms a political and mobilizing power that can never be neglected." Sandercock et. al (1998, p. 199) argue that "an everyday politics of dialogue and negotiation as the habit of political participation is foundational for a just society".

In addition, as a spatial-temporal concept,social justice has diverse meanings, but the contextualization of social justice can help us better understand its meaning.Fainstein (2005) highlightsthat aside from planning theories, the various social contexts of each specific planning should be analyzed. Fainstein (2005, p. 126) points out that"in the case of urban planning, that context is the field of powerin which the city lies", and suggeststhat as a way to understand the context of urban planning, the outcome[①] of urban planning should be investigated and compared in accordance with "a just city", which is "theorizing about justice within particular urban milieu". In terms of evaluation of the concept "just city", Sanderock understands just city to be "socially inclusive", which highlights that the identity, diversity and needs from different groups, especially the minorities should be included into the consideration of planning decision making system. Based on the above discussions, process, practices in everyday pattern, contextualization and inclusiveness, etc. are specific keywords to understand social justice.

3.1.2 Social cohesion

Modernist planning which adopts scientific rationality and technical approach to achieve physical progress has been criticized to have no contributions to the social cohesion and ever enlarging inequalities (Sandercock, 1998).In a neighborhood or community level, social cohesion means "harmonious interactions and mutual support amongresidents, and is integral to the social sustainability of the neighborhood and results in residents' satisfaction with life...Upholding social sustainability, through such social tenetsas cohesion, is a core component of vitalization efforts in poorneighborhoods"(Cheung & Leung, 2011, p. 564). In areas of social sustainability, social cohesion is highlighted as a significant aspect of the quality of societies (Berger-Schmitt, 2000) and particularly in a community level, social cohesion is pursued as most important value which could benefit community identity and bring people together (Uzzell, Pol, & Badenas, 2002). Accordingly, in this study, community social cohesion is identified as a key indicator of urban social sustainability, and the qualitative social capital based on meaningful social relationship, for example trust and norms, etc. which facilitate to construct community social

---

① The outcomes examines "each groups' ability to access employment and culture, to live in a decent home and suitable living environment, to obtain a satisfying education, to maintain personal security, and to participate in urban governance" (Fainstein, 2005, p. 126).

value——— community social cohesion should be examined.

Social cohesion as a broad concept has many constructs. For example, closeness, network density, reciprocation of resources, trust, volunteerism, sharing, etc. are all discussed as indicators of social cohesion (Cheung & Leung, 2011, p. 564). Nevertheless, with the enlarged perspectives, various influential attitudes and behaviors were added as indicators for social cohesion factor analysis and measurement. According to Friedkin (2004, p. 413), variables of social cohesion could be defined by different ways, hence " no single constructs is labeled as the basis of social cohesion". Therefore, instead of focusing on social cohesion measurements, social cohesion shall focus on the interrelationships among different social cohesion constructs. Accordingly the identification of social values which are normally gained from continuous social interactions within a specific society/community, and the dynamic interactions of different social cohesion constructs could be keywords to understand social cohesion.

As discussed above, urban social sustainability is a complex concept which involves many aspects. In sum, it is about a "long term survival of a viable urban social unit" (Yiftachel & Hedgcock, 1993, p.140). However, its significance is not equally addressed as environmental and economic sustainability. In a way, social sustainability is particularly important in urban conservation because, compared with new developments, old urban quarters contain social relationships and contexts that are often more complex. In addition, their social values are normally constructed from the long-term stable social relationship and everyday interactions, which make it more fragile to hasty changes from outside. Thus social sustainability will constantly challenge urban planners. Therefore, it is essential that in the process of urban conservation, urban planners shall adopt certain methods and framework for urban social sustainability.

Putnam (1995) points out that social organizations, norms and social trust are essential to social sustainability, and argues that public policy makers should pay more attention on the formation of social capital.In an effort to strengthen the social dimension in sustainable development, Cuthill (2010, p. 366) also argues that "social capital provides a theoretical starting point for social sustainability". Accordingly understanding social capital theory and the adaptation are important in building up social sustainability in urban conservation.

3.2 Social capital

3.2.1 Defining social capital

The definitions of social capital vary, and many scholars and institutions have defined social capital based on diverse perspectives, among which Bourdieu (1986), Coleman (1994), and

Putnam (1993) are the prominent theorists. Bourdieu (1986, p. 249) defines social capital as "the aggregate of the actual or potential resources which are linked to possession of a durable network of more or less institutionalized relationships of mutual acquaintance and recognition". Bourdieu views social capital as the asset of membership-based relationship network, especially among social elites, the use of which helps members gain privileges and profits; subsequently, he defines social capital as "the sum of resources, actual or virtual, that accrue to an individual or a group by virtue of possessing a durable network of more or less institutionalized relationships of mutual acquaintance and recognition" (Bourdieu & Wacquant, 1992, p. 119). According to him, social capital, together with economic capital and cultural capital comprise the "three forms of capital". Bourdieu's scope of social capital is focused on the individual level, and an individual's social capital account is decided by the number of connections within his relationship networks that can be used and the degree of capital that these networks possess(Field, 2008).

Unlike Bourdieu who is interested in social inequality and the membership-based privilege of the social elites, Coleman's interest in social capital lies in how social relationships can influence the educational achievements and social inequalities of students. Coleman (1994, p. 300) defined social capital as "the set of resources that inhere in family relations and in community social organization and that are useful for the cognitive or social development of a child or young person. These resources differ for different persons and can constitute an important advantage for children and adolescents in the development of their human capital " (Coleman, 1994, p. 300). Coleman's important contribution to social capital helps to explain social capital from the perspective of function: "Social capital is defined by its function. It is not a single entity, but a variety of different entities having two characteristics in common: they all consist of some aspect of a social structure, and they facilitate certain actions of individuals who are within the structure" (Coleman, 1994, p. 302).

Putnam's contribution to social capital is in terms of public policy and civic engagement. He defines social capital as: "features of social organization, such as trust, norms, and networksthat can improve the efficiency of society by facilitating coordinated actions" (Robert D. Putnam et al., 1993, p. 167). Using the metaphor of Bowling Alone, Putnam portrays a vivid image of American society's declining social capital: people are seldom engaged together in public activities, which causes the loosening of social bonds. Putnam attributes four major reasons for the decline: "urban mobility", "urban sprawl", "home–based entertainments" and "generational change", among which urban mobility and urban sprawl are the most influential factors (Field, 2008).

Bourdier, Colemen and Putnam have different viewpoints in defining social capital. Bourdieu considers social capital as the assets of the elites or people at the upper end of the class strata (Field, 2008), and he focuses on how elites use social capital to enhance their ruling powers and, hence, reproduce inequality (S. Wong, 2008). Coleman holds the view that social capital could also be the assets of lower class people. While Bourdieu (1986) regards social capital as an individual resource, Coleman extends it to the group level. According to Putnam's understanding, social capital can be shared on a societal level (Field, 2008). His study focuses on the civic engagement and associational life(Robert D. Putnam et al., 1993), and he argues that associational life is an important foundation of social order (Field, 2008). As such, Putnam's study is fixed largely on the community level.

Lin studies social capital from the perspective of social network resources. Lin defines social capital as "… resources embedded in a social structure which are accessed and/or mobilized in purposive action" (Lin, 2002, p. 41). Social networks and relationships are regarded as resources. Lin(2002) highlights the conditions of social capital: the change, difference and production of social capital and proposes seven propositions on the theory of social capital and individual action.

Fukuyama defines social capital as "an instantiated informal norm that promotes cooperation between two or more individuals" (Fukuyama & Institute, 2000, p. 1). To him, even though social capital is mostly derived from "religion, tradition, shared historical experience and other factors that lie outside the control of any government", government can contribute to the creation of social capital through education and "efficiently providing necessary public goods, particularly property rights and public safety" (Fukuyama & Institute, 2000, pp. 15,18).

World Bank's definition clearly links social capital with sustainable development. According to World Bank (2000, p. 9): "Social capital refers to the institutions, relationships, and norms that shape the quality and quantity of a society's social interactions. Increasing evidence shows that social cohesion is critical for societies to prosper economically and for development to be sustainable. Social capital is not just the sum of the institutions which underpin a society——it is the glue that holds them together."

3.2.2 The nature of social capital

1) *Public good and collective action*
According to Bourdieu, social capital is regarded as a resource that an individual owns. Coleman extended the profit of social capital from individual level to group level. According to Coleman, rather than private good, such as physical capital or human capital, social capital as a public good——"social structural resources" (Coleman, 1994)——will benefit all the members

within the group, instead of only paying off those who invest efforts. "People often might be better off if they cooperate, with each doing her share"(Pena, Lindo-Fuentes, & Unit, 1998). This explains why cooperation and collective action exist under the assumption in rational choice sociology that each individual only seeks his own optimum profit over others.

2) Trust and norms of reciprocity

Coleman identified obligation and expectation, trustworthiness, informal network, norms and sanction as main forms of social capital (Coleman, 1988). In general, social capital has three important features: trust, norms of reciprocity, and networks of civic engagement (R. Putnam, 2000, p. 288). Social capital promotes trust and norms (Coleman, 1988; Fukuyama, 1996).To Fukuyama, trust means "the expectation that arises within a community of regular, honest, and cooperative behavior, based on commonly shared norms, on the part of other members of the community" (Fukuyama, 1996).

"Social capital greases the wheels that allow communities to advance smoothly. Where people are trusting and trustworthy, and where they are subject to repeated interactions with fellow citizens, everyday business and social transactions are less costly"(R. Putnam, 2000, p. 288). Rather than one-time game, social capital, with its features of reciprocity and trust is usually produced in a process of repeated game within a stable network of community (Robert D. Putnam et al., 1993). Associations in the community level need people to cooperate. Within this process of repeated game, there emerges the norms of reciprocity and trust, as well as obligation and expectation, which can resolve the dilemma of collective action under the logic of rational choice, reduce transaction costs, and increase the efficiency of governance. Norms of reciprocity, in turn transform personal trust to social trust which is "not only internalized——through a process of socialization, but are also enforced through mechanisms which establish sanctions and third-party enforcement for violations" (Pena et al., 1998). In general, the dense network of associations increases the chances of repeated game, which promotes trust and norms (Pena et al., 1998).

3) Social capital interplays with other forms of capitals

According to Bourdieu (1986), social capital, together with economic capital and cultural capital composes the "three forms of capital". Lin (2002) pointed out the that people also invest social relations into marketplace and gain returns. According to Berge, the market can be "economic, political, labor, community, etc" (Berge, 2007). Social capital interplays with economical capital and cultural capital (Bourdieu, 1986). The accumulation and maintenance of social capital requires individual and collective investment in the group's social network relationship and the investment of economic and cultural capital, and in the long run the profit will convert to economic and cultural capital (Bourdieu, 1986). Coleman's study on the relationship between social inequality and student's educational achievement in school indicates that peer group, family and community have significant influence. He identified social capital's contribution on human capital's development and their interconnection.

4) *Scope of social capital*
Coleman (1994) highlighted social capital in three different scopes: micro-level social structure, meso-level and macro-level social structure. In the study of social capital and economic development, Woolcock (1998) also examined the "embedded" and "autonomous" social ties in both micro and macro level. In micro level, embeddedness means "intra-community ties" (integration) and autonomy means "extra-community network" (linkage); while on the macro level, embeddedness means "state-society relations" (synergy) and autonomy represents "institutional capacity and credibility" (Organizational Integrity).

3.2.3 Forms of social capital: bonding, bridging and linking social capital

It has been acknowleged that it is useful to investage the multiple dimensions of social relations and identities at the community level through three forms of social captial, which are bonding, bridging, and linking social captial (Gittell & Vidal, 1998; Wakefield & Poland, 2005). According to World Development Report, 2000/2001 "Attacking Poverty" (2000), bonding social capital is "the strong ties connecting family members, neighbors, close friends, and business associates"(Bank, 2000). Usually bonding social capital is shared by people with similar demographic characteristics. Bridging social capital is "the weak ties connecting individuals from different ethnic and occupational backgrounds" (Bank, 2000). While bridging social capital features the horizontal network structure, linking social capital is characterized as "vertical ties between poor people and people in positions of influence in formal organizations, banks, agricultural extension offices, the police" (Bank, 2000, p. 128).

Although the rich stock of social capital can contribute trust and norm, social capital does have the character of social exclusion caused by the form of bonding social capital (Blokland, Blokland, & Savage, 2008; R. Putnam, 2000; John Scott, 1991), which is to "reinforce exclusive identities and maintain homogeneity"(J. Field, 2003, p. 32).

Generally, bonding social capital is "the strong ties connecting family members, neighbors, close friends, and business associates"(Bank, 2000, p. 202). Bonding social capital could generate a high level of solidarity inside of the group; for a common purpose, it could mobilize units and resources in social groups (Galston, 2001; Narayan & Cassidy, 2001). According to Woolcock (2001), bonding social capital could be considered as a foundation form in the establishment of bridging and linking with other groups. However, some researchers have shown that bonding social capital also has its own negative functions. Based on the discussion of Portes and Landolt (Portes, 2000; Portes & Landolt, 1996), the negative sides of social capital are generally associated with bonding social capital.Usually, high degree of homogeneity could be considered as a common characteristic for the groups defined by bonding social capital (John Field, 2003; Grootaert, 2004; R.D. Putnam, 2002).

Bridging is another form of social capital defined by Putnam, which can "bring together people across diverse and social divisions" (De Roest & Noordegraaf, 2009, p. 216) and "generate broader identities and reciprocity"(R. Putnam, 2000). Putnam's forms of social capital investigate the exclusiveness and inclusiveness of social capital, which however, overlook the vertical ties which link groups within a hierarchy (Ozmete, 2011). This term is normally used to predict "more heterogeneous horizontal social networks that give people access to valuable resources and information outside their immediate network of friends and relations" (Babaei, Ahmad, & Gill, 2012, p. 2641).

The third form of social captial is linking social captial, which examines the group's relationship network with outside resources and may influence their welfare dominantly. It is formed by "vertical connections that connect individuals and groups with institutions" (Babaei et al., 2012). According to Pretty (2003), through linking social captial, groups are able to interact with other insitutions and change their polices and acquire resources. Besides the function to relate individuals and groups in diifferent social status and wealth (Côté & Healy, 2001), Woolock (2001, p. 13) addressed that linking social capital also has the capacity to "leverage resource, ideas and information from formal institutions beyond the community." In general, linking social captial is critical to distribute resource and exachange ideas and infromations; consequently it might play a very important role to achieve social well-being.

3.2.4 Social network structure

As one of the key elements of social capital (Coleman, 1994; Lin, 2002; Putman, 2000), social network is important for people to access existing social capital and to generate new social capital. Putnam indicated that "the core idea of social capital theory is that social networks have value…social contacts affect the productivity of individuals and groups" (Putman, 2000, p. 19). It is networks of associations that encourage collective collaboration (J. Field, 2003; Ozmete, 2011; Robert D. Putnam et al., 1993) and generate the "norms of reciprocity and trustworthiness" (Putman, 2000, p. 19). "People connect through a series of networks and they tend to share common values with other members of these network; to the extent that these networks constitute a resource, they can be seen as forming a kind of capital" (J. Field, 2003, p. 1).

From an organizational perspective, Nahapiet & Ghoshal (1998, p. 228) categorized social capital in three dimensions to investigate social capital's function in generating intellectual capital: "the structural, the relational, and the cognitive dimensions". The structural social capital explores the features on "network ties", "network configuration", and "appropriable organization"; the relational dimension examines the "trust", "norms", "obligation", "identification" ; while the cognitive dimension investigate the features on "shared codes and language", "shared narratives" (Nahapiet & Ghoshal, 1998, p. 239). Many features of the three dimensions are "highly interrelated", hence Nahapiet & Ghoshal (1998) assert that the aim of their analysis is "to indicate the important facets of social capital rather than review such facets exhaustively" (Nahapiet & Ghoshal, 1998, p. 228). The categorization of the structural dimension of social capital contributes to the further understanding of social capital, such as network configuration and network linkages in terms of organization network (Alguezaui & Filieri, 2010).

In this regard, to understand social network configuration, both density and components should be examined. Firstly, the configuration of network reveals the extent of network connection. Coleman proposes the model of cohesive network in which all members are linked within the social relationship network and argues that a dense, well-connected network will benefit in exchanges of information and resources, trust, norms of reciprocity, and collective action, etc. (Coleman, 1988). In terms of economic development, Putnam (1991) claims that in well-connected societies the whole economic situation is better than that in the poorly-connected societies. Well-connected networks also contribute to education, health and crime control (J. Field, 2003; Wallin, 2007). In terms of intellectual innovation, close social network has also been demonstrated to advance a firm's performance and innovations (Ahuja, 2000; Landry, Amara, & Lamari, 2002).

Nevertheless there still exists a debate on the advances regarding the configuration of social network, namely sparse network structure and dense network structure (Alguezaui & Filieri, 2010). In a sparse network structure, members are less well-connected and, hence, constitute holes within the structure. By filling the linkage gaps among members, brokerages normally lead to creativity and advances in competitions (R.S. Burt, 1995; Ronald S Burt, 2004). On the other hand, in a densely connected network, members tend to be reluctant to outside members relationships, thus causing the network to close inwardly to a certain extent. The closure of a network structure, according to Uzzi (1997), could result in a lock-in effect in the network structure, which intends to maintain the dense network and strong relationship ties and tends to reject outside relationships, information, knowledge, resources, or development opportunities (Alguezaui & Filieri, 2010).

Secondly, the configuration of network also examines the components of social network. Putnam, et al. (1993, p. 173) identify two ideal forms of social network structures: "horizontal", which is "bringing together agents of equivalent status and power"; and "vertical", which is "linking unequal agents in asymmetric relations of hierarchy and dependence". In their comparative study on the institutional development in Italy, social capital was adopted to explain the different performance of public policy and civic engagement between the northern region and southern regions. Here, the significance of horizontal network structure in promoting social capital building and reciprocity was highlighted (Robert D. Putnam et al., 1993).

3.2.5 The adaptation of social capital theory

The research scope of social capital has extended from sociology and politics to many other fields, including urban conservation. The following four areas provide insights for the adaptation of a social capital approach to urban conservation.

1) The means of "attacking poverty" and inequality

Woolcock (1998) links social capital with poverty and social economic development in underdeveloped areas. To him, a rich storage of social capital will secure an advantaged position when faced with poverty and economic vulnerability. World Development Report, 2000/2001 (2000) advocates that building up social capital for poor people is an important approach to alleviating poverty. This report argues that three ties, namely bonding, bridging, and linking social capital. Bonding and bridging capital is important for a community to secure basic needs, while linking social capital and external support, such as NGOs, are important for economic opportunity and long-term development (Bank, 2000). However, given that governments only provide limited driving forces for development, the informal network within and between poor communities are significant, as it is based on trust and cooperation. Bourdieu(1986)was concerned about the production and reproduction of inequity through the combination of economic capital with social and cultural capital and saw social capital as an "asset of the privileged and a means of maintaining their superiority (Field, 2008). According to Coleman (1994), social capital as a resource, is not limited to social elites. It is also an asset of poor people which secures their jobs and basic needs through informal networks of information and labor market. As an asset of community, it also contributes to the community resources (Berge, 2007).

2) Solving economic development dilemmas

Woolcock (1998, p. 171) discussed the functions and forms of social capital in both bottom-up and top-down development dilemmas. According to Woolcock, at the micro level through which bottom-up development usually emerges, intra-community networks are important. While excessive bonding social capital will undermine economic development. To solve this dilemma, this study argues that "linkage to broader extra-community institutions" should complement bonding social capital in order to initiate further development (Woolcock, 1998). In a macro level top-down development, the dilemma lies in that different "combinations of the state's organizational capacity and engagement with and responsiveness to civil society" will lead to "collapsed state", "weak states" or "developmental states" (Woolcock, 1998, p. 176). The dynamic interaction between "top-down"and "bottom-up" development is highlighted as Woolcock (1998, p. 180) argues that"successful development potentially sets in motion an underlying dynamic creating new distributional coalitions with fresh claims to make on the resources of both the state and other social groups", which can be understood as linking social capital.

Woolcock's study highlights the significance of different forms of social interactions which happen both at the micro-level, macro-level and inbetween. Rather than focusing on the outcomes, it investigates the combinations of different social capitals at different levels from the perspective of "history and institution process", which proves that social relationships contribute greatly in building up civic life and economic devleopment, and especially bridges the gap that "an institutonal foundations" not only exist in macro-level development but also exist in micro-level development (Woolcock, 1998).

3) *The field of community governance*
The research by Putnam et al. (1993) in Italy indicates that the storage and distribution of social capital will influence people's participation on public policy and their attitudes towards institutions. A rich storage of social capital that is evenly distributed usually generates a strong sense of community and a willingness to participate, which in turn influences the efficiency of community governance. The dimensions of social capital (bonding/bridging/linking social capital) and their cooperation theory are also adapted in community governance. The cooperation of the three ties of relationship network will increase the participation of community and prevent segregation of members and injustice of decisions. For example, if the vertical network is very loose or broken, issues of justice will arise.

4) *Institutions and organizations and policy making*
Fourthly, the concept of social capital has been linked to the field of institutions and organizations and policy making. From the perspective of an organization, Cohen et al. (2001) presents social capital as the "raw material of positive relationship" between people——trust, understanding, shared value and collective action——which will promote cooperation within organizations. Stiglitz (2000) advocates an understanding of social capital based on an organizational level. He discusses the relationship of formal and informal institutions and argues that, as a society progresses along with economic development, informal institutions based on personal or

community interaction will to some extent be replaced by formal institutions based on a market economy. Stoglitz asserts that this superseding of informal institutions by formal institutions undervalues the importance of informal networks at the individual and community levels, as it overlooks the cooperation among different dimensions of social capital and, hence, is frequently attributed to the cause of injustice. Given that government provides limited driving forces for development, the informal network within and between poor communities is significant, as such a network can help poor communities become self-supportive through trust and cooperation.

Woolcock (2001) conducted a multi-dimensional approach to investigate the diverse combinations of the three forms of social capital and their outcomes in social and economic development, which indicates that the dynamic relationship of network-based institutions, organizations, and associations is a significant direction that policy making should adopt. The combination with its aims to gain "optimal" rather than "maximum" condition is featured as a process of changing, adapting, and negotiating. For example, when formal institutions are absent or weak, informal associations will emerge from the bottom-up to function. Francis Fukuyama (2000) also highlighted the informal form of interaction in a micro scale. Woolcock appealed for the adaptation of social capital in public policy based on a broader institutional context.

In addition to the above areas, topics of "good will" among groups (Adler & Kwon, 2002), investment and income (Glaeser, Laibson, & Sacerdote, 2002), dilemma of collective action, game-theory and economic efficiency (Ostrom, 2000) are discussed.

3.2.6 The potentials to adapt social capital approach in urban conservation decision making

Based on the above discussions, there existpotentials in adapting a social capital approach in urban conservation under a decision-making framework of path dependency, especially in order to enhance social sustainability process.On one hand, social justice and social cohesion as key indicators of social sustainability could be investigated via social capital approach. Social capital highlights the process rather than the outcome (Woolcock, 2001), and its adaptationof policy helps achieve collective action and reciprocity, promoting dynamic relationships between networks. Moreover, the dynamic interactions among diverse social value constructs could also be revealed via social capital survey. On the other hand, an observation of the different roles and combinations of different forms of social capital in the social network of decision making might reveal how decision-making was influenced and retargeted in the urban conservation process. Hereby, social capital is specifically understood as resources, driving forces, and network.

In addition, as has been discussed above, social cohesion is a core social value which is composed by many constructs. Also it is more important to examine the dynamic interactions of different value constructs; it is also meaningful to understand how these social value constructs are formed based on long-term stable social relationship, especially in the traditional community. As social capital has a wide brand of definition, hereby social capital could be specifically taken as resources of social relationship which could in turn construct the stable core social value among a traditional community, for example " HE", which means social cohesion and in China.

3.3 Social relationship pattern in traditional society of China

There already exists rich literature of social capital based on the studies of modern China society, as can be seen in the study of social network and relationship ties(Bian, 1997a, 1997b; Bian, Breiger, Galaskiewicz, & Davis, 2005), GUANXI (Bian & Ang, 1997; Guthrie, 1998; Y. Wong & Leung, 2001), inequity(Logan & Bian, 1993), social capital and poverty (Zhang, Lu, & Zhang, 2007), etc. With regards to social relationship and social capital in traditional Chinese society which includes many traditional neighborhoods, the special relationship pattern: CHA-XU-GE-JU developed by Fei (1947) could provide fundamental understandings.

3.3.1 Chinese social relationship structure CHA-XU-GE-JU（差序格局）

In his book *Xiangtu Zhongguo* (1947), Fei uses ripples formed by casting a stone into water as a metaphor of the structure of traditional Chinese society.[①] Fei states that the principles to organize Chinese society are fundamentally different from the prevailing western ones, which are basically through the unit of "group pattern", like "bundles of firewood".

Fei develops the concept of *CHA-XU-GE-JU* to describe the social relationship structure in traditional Chinese society, which is the "the Pattern of Difference Sequence" (Fei, 1992) . GE-JU, means the pattern of Chinese society, which is depicted in two dimensions:"CHA"(Difference) and "XU" (Sequence) as "nonequivalent,ranked categories of social relationships"(Fei, 1992), or Difference and Sequence (Yan, 2006). "Difference" can be understood as to identify whether the relationships are closely or loosely based. "Sequence" refers to the hidden order and social hierarchyin causing these relationships differences, which according to Fei, are largely referred to the preoccupied Confucian ethics and traditional morality (Fei, 1947; Yan, 2006).

And the pattern of difference and sequence——CHA-XU, as organizational principles play profound roles in forming the basic relationship characters of traditional Chinese society, which

---

[①] Each person is located in the center of ripples, or concentric circles; and every wave circles represent the social influence of the person: as the circles spread further, the influence gets thinner.

can "provide the structure framework for social action, they are intuitive and taken for granted, they are deeply embedded in people's world views, as well as in the society that people re-create every day" (Fei, 1992, p. 19). This social structure reflects many deep-rooted impacts of traditional Chinese values and culture, and the Pattern of Difference Sequence exists in local society and community level.

According to Fei, rural Chinese society can be identified by the following characters: the importance of kinship and family; the ethic value of egocentric thoughts; therelative public-private relationship, or the relativity of ego and group; traditional social relationships, ethics and morality which maintain social order, such as "KE-JI-FU-LI" (克己复礼) and self-cultivation; as well as the patriarchal control system in political mechanism (Fei, 1947).

Among the above characters, the term " JI" (己)——" Self" in the egocentric thoughts are explored in both perspectives of social structure and personality. In the social structure aspects, a "self" is a kinship and family oriented "self", rather than the pure "individual" compared to western culture (Fei, 1947; Piao, 2003). And in the personality aspect, this "self" is dependent on the relative relationships with others and does not represent an independent personality in whole. It is the socialized process accompanied with moralization of "REN-LUN" (人伦)——virtues and principles that formulate the whole body of personality composed by the relative "JI"——"self" in diverse relationships with others (Piao, 2003).

In addition, with regards to the egocentric thoughts in a person's relationship with others, each relationship with others is relative and determined according to the relationship center——" Self " in different social interactions. According to Yan (2006), the relativity of a person's relationships with others indicates the continuous process of adjusting and redefining his social position, his roles and his value of existence in every social interaction. The relationship pattern " CHA-XU-GE-JU" leads to " CHA-XU-REN-GE" (Difference-Sequence Personality), which refers to the personality integrated with relatively nonequivalent and diverse social roles as well as social positions.

The capacity of a person to prevail in the social structure of " CHA-XU-GE-JU" means well connected social relations with others, which in turn may leads to the flexible "CHA-XU-REN-GE" (Difference-Sequence Personality) , which is resilient to adjust and redefine the personality (Yan, 2006). A good example is "SHU-REN-SHE-HUI"(熟人社会), which refers to the acquaintance society in China.

From the above exploration, it is very clear that the "CHA-XU-REN-GE", namely "Difference

Sequence Personality" based on the egocentric thoughts has largely differentiated the social relationship of traditional Chinese society with that of western society.

Moreover, although there are difference between rural society and urban society, traditional society and modern society, as well as agrarian society and industrial society, *CHA-XU-GE-JU* represents the organizational principles and basic values of Chinese society(Fei, 1992). Yan (2006) argues that as long as the ranked categories and basic values (Confucianism ethics) remain, CHUA-XU-GE-JU remains effective in formulating societies and social relationships, as well as personalities.

Yan (2006) further develops the concept of CHA-XU-GE-JU in terms of commonly accepted core values among different contexts and argues that the social relationship patterns may have different presentations due to the differences such as traditional society vs. modern society, rural vs. urban society, and agrarian society vs. industrial society, ect., it is also heavily relied on the commonly accepted core values of Chinese society. And as long as the core value remains, the organizational principles of the entire society, as well as the differential sequence pattern of personality will endure in both modern cities and rural areas. In light of this, the commonly accepted core values of Chinese society will be explained in the following session.

3.3.2　Core values of Chinese society ——HE (和 /Cohesion)

HE (和), namely "cohesion" has been widely shared as the core values of traditional Chinese society. Under the fundamental social relationship structure CHA-XU-GE-JU, the concept of "HE" could be developed in two levels. Firstly in the national level, HE means to unite, and a pattern of "Diversity in Unity" (和而不同) is formulated under the CHA-XU-GE-JU, with the cohesive core being the advanced culture, and the cohesion as a process for the entire nation to integrate around the cohesive core during the long history (Fei, 1989, pp. 16-18).

In the local society and community level, cohesion represents the core social values and could be summarized as "social cohesion and social solidarity", which are deeply rooted in the traditional Chinese culture and social value (Tan & Tan, 2014, p. 191) . It is widely accepted that social cohesion in the local society and community level are constructed by five social norms "Benevolence(仁)-Righteousness (义)-Propriety (礼)-Wisdom (智)-Faithfulness (信)". The social cohesion in this level could also be understood as a process of an individual to be closely and harmoniously integrated into the society, with the continuously adjusting of himself from the egocentric thinking of his relationship with others (Yang, 2013).

Chapter 4   Research Methodology

To answer the research questions and test research hypothesis, research methodology is developed which contain several approaches and methods. In this chapter, the research methodology employed in this study is described. In the first section, the research approach is introduced, which is case study approach. The second section describes the research methods, including qualitative and quantitative methods. Although the case study approach and the above methods might not answer research questions directly, the findings could provide significant understandings and could facilitate the following methods to answer research questions.Noticeably, to answer the specific three research questions, main research tools, the social capital survey and social network analysis under the framework of path dependency are discussed respectively in the third and forth section. Based on the findings of filed investigation, interviews, and questionnaires, social capital survey could be able to answer the first two research questions, and social network analysis could be able to answer the third research question. In the fifth section, the main steps of the research are addressed, which is followed by the summary of this chapter and research framework.

4.1   Case study approach

As Robert K. Yin defines that the case study method is "an empirical inquiry that investigates a contemporary phenomenon within its real-life contest; when the boundaries between phenomenon and contest are not clearly evident; and in which multiple sources of evidence are used"(Yin, 1984, p. 23).Case study is an effective approach to test hypothesis, and through study of selected cases, researches gain a clear understanding on "how" the development is and attain an insight on both interior and exterior driving forces. Case study approach is quite intensive on a specific context. It often combines a number of methods, such as observation, interviews, questionnaire, etc., and in this research, by employing case study approach, the followings objectives are expected to be achieved: 1) identifying heritage values; 2) identifying driving forces and affecting factors; 3) understanding the entire development path of each case.

This research chose two cases in China. One is Hui Fang in Xi'an, a provincial capital city in Western China; and the other one is Tianzi Fang in Shanghai, a well known global city.Even though the backgrounds and development contexts of the two cities are different, the two cases provide the possibility to examine the common connection, which is the social capital initiated urban conservation, as can be seen from the following common bases:

- In terms of physical and social aspects, both cases are historic urban areas as well as specific social areas; respectively residents are identified by ethnic status in case of Hui Fang, and by social economic status in case of Tianzi Fang.

- In terms of development approach, both cases are proposed by local government with a top-down approach as redevelopment project.

- In terms of development dilemma, both cases witnessed the demolition of surrounding neighborhoods by the mainstream urban development approach practiced in most Chinese cities——urban redevelopment with mass demolition and mass construction.

- In terms of transitional approach, both cases were initiated by different forms of social capital, survived physically and socially from demolition and relocation, and successfully created new paths for conservation decision-making patterns.

Given that the objectives of this research focus on revealing the common roles of social capital in urban conservation process among different urban contexts, this research will not take efforts to compare these two cities, nor will this research compare the measurement of social capital and social network between two cases.

Firstly, urban context plays a very important role in urban heritageconservation, and every case has a specifically different context. Urban contexts are complicated and cannot be duplicated. As this research focuses on how social capital facilitated the transformation of the entire conservation process, it is more necessary to investigate how urban context shape the interactions among different forms of social capital within each case, rather than comparing the urban context among different cases.

Secondly, regarding social network analysis, the interest of this research lies in the transformational process of the decision-makingpattern. Therefore the comparison between different cases which could only provide quantitative results will be spared. Just as Friedkin(2004, p. 422) argues, "research on social networks should begin to specify more clearly the social processes in networks that are affecting individuals' attitudes and behaviors. It does not suffice to assert that certain network structures foster cohesion in the absence of an explicit model of the social processes that link the network structure to individual outcomes, because similar network structures may have dramatically different implications for individual outcomes depending on the social process that is occurring."

4.2 Research methods

4.2.1 Qualitative methods

1) *Literature review*

Urban conservation and social capital have been the research focuses in urban study and sociology for a long period. The research will firstly review the published literature. Through comprehensive reading, researcher could understand, classify and sum up current research opinions and achievements and then could identify research gaps. Meanwhile the general information of cases is equally important, released researches about the cases, published governmental reports, local newspapers, literary works, event records, and oral history.

2) *Field investigation*

It has been emphasized above that case study is the main approach of this research which aims to identify heritage values, development path, the driving forces as well as affecting factors. In addition, it has been highlighted that contextualizing each conservation process is very important to understand social sustainability. Therefore, field work is essentially necessary to understand the two cases, especially in terms of the physical configurations of each case, difficult spatial functions among each community, local people's daily practices, and tourists' influence etc., aside from specific tangible values of historic buildings. Although the findings of field investigation might not answer the research question directly, they will provide essential understandings to facilitate the value definition process and to contextualize the decision making process.During field investigations, four research measures were put in use: observation, mapping, photography and video record, and collecting archival document.

- Observation

The contents of field observation include the everyday lifestyle of local residents, the activities of locals and tourists, artists. Furthermore, the relationships among residents and with outsiders are also the focus of observation.

- Mapping

Mapping is useful to understand the physical context of cases and different groups' diverse everyday practice in urban heritage. In addition, mapping of institutions, everyday activities and community facilities are also important to analyze the social relationships.

- Photography and video recording and story collection

Along with the field investigation, photography and video methods will be employed to record the activities. A record of stories telling from both old residents and new members will be conducted to gain their perceptions of change.

- Archival documents collection

Relative archival documents will be collected. Reports from government departments, census publications and statistic yearbooks, etc. will be amassed. Historical data, images and other information related to the transformation of cases will be gained through archives, local libraries and planning museums. Urban conservation planning and land use plans will be acquired from local governments such as the Planning Bureau.

3) *Interview and focus group discussion*

In order to gather the rich information about the conservation processes and the social relationships in these two cases, in-depth interview, semi-structured interview, and life stories have been conducted. Focus Group discussion and interviews with both important persons and peripheral persons were also carried out. Through the observation, some residents have been chosen as interlocutors. Additionally, some tourists will also be selected to be part of the study.

In the interview and group discussion, various relevant authorities, institutes, associations, and individuals were consulted and interviewed.The following Tables (Table 4-1 and 4-2) summarized the main interviewees who have been closely involved in the conservation process of both cases.

Main interviewees during fieldwork of Hui Fang, Xi'an   Table 4-1

Main Interviewees during fieldwork of Hui Fang, Xi'an				
Interview Period: September 2012—November 2012				
Name	Affiliation	Position	Interview Methods	Times
Mr. He	Xi'an City Planning Bureau	Former Bureau Director,In charge of SRP 2005—2007	semi-structured interview	1
Mr. Wu	Xi'an City Development Researching Design Center	Urban designer,In charge of SRP 2010—	semi-structured interview	2
Mr. Tian	Xi'an City Development Researching Design Center	Urban designer,SRP 2010—	semi-structured interview	2
Prof. Lv	Xi'an University of Architecture and Technology	Advisory Board Member of Xi'an City Planning Commission	semi-structured interview	1
Mr. Xi	Xi'an Planning and Design Research Institute	Director	semi-structured interview	2
Mr.Shi	Xi'an Planning and Design Research Institute; Xi'an Muslim Historic District Protection Project Office	Architect, HMHDPP 1997—2002	in-depth interview	5

| \multicolumn{5}{c}{Main Interviewees during fieldwork of Hui Fang, Xi'an} |

Name	Affiliation	Position	Interview Methods	Times
		Interview Period: September 2012—November 2012		
Mr. He	Xi'an Planning and Design Research Institute	Planner, Xi'an Master Plan 2004—2020	semi-structured interview	4
Ms. Liu	Xi'an Planning and Design Research Institute	Head, Conservation department	semi-structured interview	2
Prof. Xiao	Xi'an University of Architecture and Technology	University coordinator of XMHDPP 1997—2002	in-depth interview	1
Prof. Chang	Xi'an University of Architecture and Technology	Tutor of *XMHDPP 1997-2002* Student workshop	semi-structured interview	1
Prof. Hoyem	Norwegian University of Science and Technology	Project coordinator of XMHDPP 1997—2002	in-depth interview	1
Prof. Zhai	Xi'an Jiaotong University	Researcher of Hui Fang	in-depth interview	1
Ms. Liu	Beiyuanmen Street Office; Xi'an Muslim Historic District Protection Project Office	Officer, Coordinator of XMHDPP 1997—2002	semi-structured interview	1
Mr. An	The Great Mosque	Religious leader	in-depth interview	1
Mr. Ma	The Great Mosque	staff	semi-structured interview	2
Mr. Ma	Xiyangshi Street	Self-employer	semi-structured interview, life stories	1
Mr. An	No. 125, Huajue Alley,	House Owner, whose house was restored during *XMHDPP 1997—2002*	semi-structured interview, in-depth interview, life stories	2
Mr. Hua	The Great Mosque	Active resident	semi-structured interview and in-depth interview, life stories	6
Ms. Feng	Female religious school; Souvenir Shop	President; Shop Owner; Active Resident	semi-structured interview and in-depth interview, life stories	3
Mr. Ma	The West Mosque	Religious leader	in-depth interview	1
Mr. Dong	Guangming Alley Photo Studio	Shop Owner; Volunteer of SRP 2005—2007	semi-structured interview, in-depth interview, focus group discussions, life stories	8
Mr. Ma	Self-employed business	Resident Representatives of SRP 2005—2007	semi-structured interview, in-depth interview, focus group discussions, life stories	9
Mr. Liu	Employee	Resident Representatives of SRP 2005—2007	semi-structured interview, in-depth interview, focus group discussions	1
Mr. Chen	DaPiyuan Street	Self-employed arts collector, resident	semi-structured interview,	1
Ms. Wang	Sajinqiao Neighborhood Committee	Director (2008—), resident	semi-structured interview	1

Main interviewees during fieldwork of Tianzi Fang, Shanghai   Table 4-2

Main Interviewees during fieldwork of Tianzi Fang, Shanghai

Interview Period: December 2012—Feburary 2013, June 2013

Name	Affiliation	Position	Interview Methods	Times
Mr. Zhang	Self Employed Business	Former consultant of Taikang Art Road	semi-structured interview, focus group discussions, life stories	1
Mr. Zhou	Tianzi Fang Trade Union; Resident	Coordinator of house leasing in Tianzi Fang; Active Resident	semi-structured interview, in-depth interview, focus group discussions, life stories	5
Mr. Zheng	Shanghai Huaxia Creative Culture Research Institute	Director, Former party secretary of Dapuqiao Sub-district Office	semi-structured interview, in-depth interview, focus group discussions, life stories	4
Mr. Wu	Tianzi Fang Trade Union	Chairmen	semi-structured interview	1
Ms. Wang	Jianzhong Community Committee	Director	semi-structured interview	3
Mr. Le	Art Gallery in Tianzi Fang	Shop Owner, Resident	semi-structured interview	1
Ms. Le	Art Gallery in Tianzi Fang	Shop Owner, artist Resident	semi-structured interview	2
Mr. Yang	Er-Jing Alley	Resident (opposed)	semi-structured interview	1
Prof. Zheng	Tongji University	Advisory board director, Historic district and buildings' conservation Board, Shanghai Urban Planning Committee	interview	1
Prof. Sha	Tongji University	Scholar and Professional of Historic district and buildings' conservation	interview	1
Prof. Liu	Tongji University	Planner Sinan Road Mansion Conservation Plan	semi-structured interview	1
Prof. Yu	Fudan University	Consultant of resident gathering	interview	1
Mr. Zhou	Tongji University	PhD, researcher of Tianzi Fang	semi-structured interview	2

4.2.2 Quantitative methods

1) *Questionnaire*:

Quantitative data is easier to collect and compare. Quantitative data can be objective and "are especially useful for determining the impact of projects and policies"(Jones & Woolcock, 2007, p. 9). The quantitative study on social capital has developed a series of framework and standard questionnaire with tested validity and reliability, which has been widely used.[①] In addition,

---

[①] For example, "Measuring Social Capital: an integrated questionnaire of the World Bank"(Christiaan, Narayan, Jones, & Woolcock, 2004, p. 5), based on six dimensions: "*Groups and Networks*", "*Trust and Solidarity*", "*Collective Action and Cooperation*", "*Information and Communication*", "*Social Cohesion and Inclusion*", "*Empowerment and Political Action*", provide a measuring instrument based on a household survey.

other quantitative questionnaire regarding the changes in economy, rent and daily life expenses are also well developed.

In this study, two rounds questionnaires were conducted in both cases. The first round aimed to investigate the community profile, which includes demographic profile, socio-economic profile, built environment and facility profile, change of residents' life in recent 5 years and their feedback, residents' attitudes towards future redevelopment, residents' perception of culture, tradition and place character, cognitive social capital, etc.; The second round of questionnaire is to conduct social capital survey within neighborhoods of the two cases, which will be specifically explained in the next sector: social capital survey.

2) Data collection of community profile questionnaire:
In Hui Fang, questionnaires were sent to two neighborhoods, namely the Great Mosque neighborhood and Sajinqiao neighborhood, and the sample size was 100 per neighborhood. The valid respondents in the Great Mosque neighborhood were 91, and in Sajinqiao neighborhood 87 respondents were valid.

In Tianzi Fang, questionnaires were sent to both remaining residents and moving-out residents. The questionnaire of remaining residents was conductedby the researcher from 16 December 2012 to 28 December 2012, and the questionnaire of the residents who moved out was conducted later by means of attending their reunion. 50 questionnaires were sent to the remaining residents and 41 were valid. 90 questionnaires were sent to the residents who moved out and 73 are valid.

4.3 Social capital survey

The hypothesis states that social capital plays an important role in social cohesion and influences the outcome and process of urban conservation. Accordingly the development of methods will focus on testing hypothesis, by employing social capital measurement approach.

Social capital needs to be both deeply and widely understood and examined. According to Krishna (1999, p. 121), in a community and individual level, social capital is measured in two dimensions: cognitive social capital and structural social capital. According to Krishna et, al.(1999, p. 10), a cognitive social refers to trust, norms of reciprocity, solidarity, beliefs, and attitudes which *"create the conditions under which communities can work together for a common good "*. Structural social capital means *"the composition and practices of local level institutions, both formal and informal, which serve as instruments of community development"*, which is normally manifested as horizontal networks. A qualitative approach will reveal the cognitive social capital and provide deep insights on the nature of community collective action and cooperation. Researches on the qualitative survey of social capital are conducted by many scholars, such as Pena, et al. (Pena & Lindo-Fuentes, 1998), Krishna, et al. (1999),Jones, et al.(Jones & Woolcock, 2007), among which, Social Capital Assessment Tool (Krishna & Shrader, 1999), provides a qualitative study with a pilot test from community, household and organizations perspective for World Bank. Using Mixed Methods to Assess: Social Capital in Low Income Countries: A Practical Guide(Jones & Woolcock, 2007, p. 3) also provide an effective reference based on six dimensions: *"Groups and Networks","Trust and Solidarity","Collective Action and Cooperation","Information and Communication","Social Cohesion and Inclusion","Empowerment and Political Action"*.

4.3.1 Survey design

In this research, social capital measurement was conducted on the neighborhood level. In the micro level, a survey of social capital was conducted based on random sampling in order to measure the community cognitive social capital and their structural social capital. According to literature review, five main variables were identified as the core indicators of cognitive social capital, namely Benevolence（仁）, Righteousness（义）, Propriety（礼）, Wisdom（智）, and Faithfulness（信）, representing the core values and qualities of the social relationships of people in the traditional Chinese neighborhoods.

Based on literature review, Benevolence, Righteousness, Propriety, Wisdom, and Faithfulness, were identified as the five main indicators of cognitive social capital in traditional Chinese neighborhood. Although there exists no particular tools to measure these five indicators of cognitive social capital currently, scholars have developed quite thorough understandings on how to reinterpret these core social values, including the connotationand the extension (Deng, 2006a, 2006b; Liu, 2010). In light of this, the questions developed for each indicator in the measurement survey are based on three levels, forming the structure of intrinsic values——principles——extrinsic values. This structure aimed to explore the intrinsic value can shape the extrinsic values among a neighborhood via principles of daily interactions.

For the first indicator——Benevolence（仁）, the explanatory questions are developed as Table 4-3.

	Social cohesion indicator of Benevolence			Table 4-3
Indicators	Levels		Do you agree the following statements?	Code
Benevolence（仁）	Intrinsic Value:	R1	Most people in this neighborhood generally care and concern about each other	1-5

Indicators	Levels		Do you agree the following statements?	Code
Benevo-lence (仁)	Principle:	R2	Normally when you are facing the bad conditions, do you want to drag the rest of this neighborhood down together?	1-5
	Extrinsic Values:	R3	Most people in this neighborhood are willing to help you if you need it	1-5
		R4	People in this neighborhood are generally peaceful	1-5
		R5	The relationships among people in this community are generally harmonious	1-5
		R6	During the last 5-10 year, the level of social cohesion in this neighborhood has (from largely decreased to largely improved)	1-5

For the second indicator——Righteousness (义), the explanatory questions are developed as Table 4-4:

Social cohesion indicator of Righteousness   Table 4-4

Indicators	Levels		Do you agree the following statements?	Code
Righteousness (义)	Intrinsic Value :	Y1	Nearly everyone in this neighborhood wants to live a life of righteousness	1-5
	Principle:	Y2	Everyone in this neighborhood deserves just treatments	1-5
		Y3	People in this neighborhood care about the just treatment of the entire neighborhood	1-5
	Extrinsic Values:	Y4	Most people in this neighborhood are willing to contribute time or money for common goals?	1-5
		Y5	People here look out mainly for the welfare of their own families and they are not much concerned with the entire neighborhood's welfare	1-5
		Y6	During the last 5-10 year, the contributions of people from this neighborhood for neighborhood common good have (from largely decreased to largely improved)	1-5

For the third indicator—Propriety (礼), the explanatory questions are developed as Table 4-5:

Social cohesion indicator of Propriety   Table 4-5

Indicators	Levels		Do you agree the following statements?	Code
Propriety (礼)	Intrinsic Value :	L1	Most people in this neighborhood humble themselves	1-5
	Principle:	L2	Most people in this neighborhood respect others	1-5
		L3	Most people pay a lot of attention to the common norms in the neighborhood	1-5
	Extrinsic Values:	L4	I feel accepted as a member of this neighborhood	1-5
		L5	Suppose residents in your neighborhood had a fairly serious dispute with each other. Who do you think would primarily help resolve the dispute?	1-5
		L6	Do you think over the last few years the function of norms and virtues have been (from largely decreased to largely improved)?	1-5

For the fourth indicator——Wisdom(智), the explanatory questions are developed as Table 4-6:

Social cohesion indicator of Wisdom　　　　Table 4-6

Indicators	Levels		Do you agree the following statements?	Code
Wisdom (智)	Intrinsic Value:	Z1	Most people in this neighborhood respect knowledge and value talents	1-5
	Principle:	Z2	Most people in this neighborhood are open-minded and want to keep pace with the times	1-5
		Z3	Most people in this neighborhood is flexible	1-5
	Extrinsic Values:	Z4	Many creative ideas have been generated in this neighborhood	1-5
		Z5	Most people in the neighborhood allow diversity in their daily lives	1-5
		Z6	During the last 5-10 to ten years, the level of prosperity in this neighborhood has (from largely decreased to largely improved)	1-5

For the fifth indicator——Faithfulness (信), the explanatory questions are developed as Table 4-7:

Social cohesion indicator of Faithfulness　　　　Table 4-7

Indicators	Levels		Do you agree the following statements?	Code
Faithfulness (信)	Intrinsic Value:	X1	Most people in this neighborhood are basically honest	1-5
	Principle:	X2	In your neighborhood people generally trust one another in matters of lending and borrowing	1-5
		X3	If I have a problem, there is always someone to help me in this neighborhood	1-5
	Extrinsic Values:	X4	Residents of this neighborhood are more trustworthy than others	1-5
		X5	One has to be alert or someone is likely to take advantage of you in collective actions	1-5
		X6	Over the last few years this level of trust has been(from largely decreased to largely improved)	1-5

As the research mainly aims to reveal that the core values of cognitive social capital as a common values system in traditional Chinese society, the study of the interrelationship between cognitive social capital and the case by case structural social capital is to facilitate the further understanding of cognitive social capital. Accordingly four variables were identified to explain structural social capital in terms of network integration, information resource, participation of formal organization and informal organizations.

For the structural social capital, the explanatory questions are developed as followings:

Structural social capital as cohesion indicator   Table 4-8

Network integration	S1	Please rank the extend of your social network closure as follows:	Code 1-5
		All my contacts are within this community	1
		Most of my contacts are within this community	2
		Half of my contacts are within this community and the other half are outside of this community	3
		Most of my contacts are outside of this community	4
		All my contacts are from outside of this community	5
Resource network	S2	Who do you think normally provide useful information such as job opportunities and finance to you via informal contacts?	Code 1-5
		Family members	1
		Close friends	2
		Neighbors	3
		Normal friends	4
		Business partners outside of the neighborhood	5
Participation of formal organizations	S3	How often do you or your family members regularly join any formal organizations in this community?	Code 1-5
		Never	1
		Seldom	2
		Now and then	3
		Quite regularly	4
		Very frequently and active	5
Participation of Informal groups	S4	How often do you join in informal groups within this neighborhood in your social network?	Code 1-5
		Never	1
		Seldom	2
		Now and then	3
		Quite regularly	4
		Very frequently and active	5

In the end, local residents' general attitudes toward the conservation of their neighborhood are investigated as a special indicator through the general question developed as followings:

Local residents' general attitudes of conservation as a special indicator   Table 4-9

Conservation Attitude	CA	Do you think whether it is necessary to conserve this community?	Code 1-5
		Strongly not necessary	1
		Not necessary	2
		No	3
		Necessary	4
		Strongly necessary	5

4.3.2 Data collection, data entry and data editing

In Hui Fang, the sample size of social capital survey was 65, among which 57 was recognized as valid. In Tianzi Fang, the sample size was 50, and the valid respondents were 43. IBM SPSS Statistics 22 is adopted to describe and analyze the social capital survey data and examine the hypothesis 1 and 2. Basically the social capital survey data processing includes three steps.

The first step focuses on the input of the survey data into the SPSS to get an optional database ready for analysis, which includes the following steps: data entry and data editing (Miller & Acton, 2009).

In the SPSS data file, each questionnaire is recorded in one row, namely one respondent, and the response to each question is recorded in each column, representing a variable. After the data entry, the variables can be defined by adding meaning and information to specific variables. For example, the questions in cognitive social capital survey normally constitute 5 answers, and in data entry five codes ranging from 1 to 5 represent each specific answer. The codes are normally categorized as——1= "Strongly disagree"; 2= "Disagree"; 3= "Neither disagree nor agree"; 4= "Agree"; 5= "Strongly agree". All these information need to be recorded to the Values Labels so that the value of different variables represents different meaning; In addition, if a respondent fails to answer a specific question or the question is not "not applicable" to the respondent(Miller & Acton, 2009), in SPSS it will be treated as missing value, normally in the column of "Missing Values" a specific code will be used to signify the "missing" values (Kremelberg, 2010; Miller & Acton, 2009). In this survey, code 9 is used to represent the missing value of variables. After the data entry, data editing is necessary to validate the data. In this step, firstly the codes entered in the variables are checked; secondly, 4 respondents from outside of these two communities have been removed in order to get a valid database for the resident's survey of these two communities.

The second step is to explore the social capital data, specifically to examine the correlation coefficient between the variables within each core cognitive indicators. Correlation coefficient facilitate to test "whether there is a real relationship between two interval/ratio variables" (Miller & Acton, 2009, p. 204), with a correlation coefficient of +1 indicating a "perfect positive relationship", with that of -1 meaning "a perfect negative relationship" and that of 0 showing "no relationship" between two variables(Kremelberg, 2010, p. 121; Miller & Acton, 2009, p. 204). Among the several correlation coefficient analysis methods, Pearson's correlation coefficient conducts "a parametric test" (Miller & Acton, 2009, p. 204)and provides inference about the parameters distribution. Accordingly Pearson's correlation coefficient has been adopted to illustrate the relationship of the six explorative variables to each other within each social capital cognitive indicator, as well as the positive or negation directions of the correlation.

The third step: data deduction of different indicators of community social cohesion, which hereby mainly refers to 5 cognitive social capital indicators (*R-Y-L-Z-X*), given that the 4 structural social capital indicators (*S1-S4*)) are simple indicators.

To analyze the correlation of the nine indicators of social cohesion, the variables within each main indicator shall be simplified into one variable or few variables. As it is believed that "the variation observed in a variety of individual variables reflects the patterns of a smaller number of some deeper, more fundamental features and factors" (Miller & Acton, 2009, p. 240), factor analysis is employed as a data deduction method to facilitate this data simplification process. Though there are many methods in factor analysis, specifically Principal Component Analysis (PCA) is adopted as it facilitates exploratory factor analysis, which means the "number of factors have not been predetermined" compared to the confirmatory factor analysis (Kremelberg, 2010, p. 288). By employing Principle Component Analysis, the six correlated variables will be converted to linearly values, among which the first principal component weighs highest in the values and has highest variability among the main indicator, and the key variables to each main cognitive social capital indicator will be explored.

After the factor analysis, the factors are identified and the nine constructs (RYLZX,S1-4) of social cohesion will have new scales on the neighborhood level by adding the mean scores of every respondent's scores on the identified factors of each variable. And the new neighborhood social cohesion constructs are now available for further analysis.

4.3.3 Examining research hypothesis 1 and hypothesis 2

1) *Examining research hypothesis 1*

In the first hypothesis, it is hypothesized that in traditional Chinese community, there exist certain indicators of cognitive (community) social capital, which are commonly important in maintaining neighborhood social cohesion, despite the diverse social structures and groups of residents.

Since the cognitive social capital——Benevolence, Righteousness, Propriety, Wisdom, and Faithfulness are assumed to be very important social cohesion indicators for social cohesion, the correlation between cognitive social capital and the structural social capital could be the main focus to test the hypothesis.

Pearson's correlation coefficient analysis of the nine social cohesion constructs is conducted to examine and compare the correlationsof cognitive social capital to structural social capital in different neighborhoods. If the factor analysis confirms the validity of cognitive social capital in the social cohesion constructs, and the reliability of cognitive social capital to structural social capital is examined, hypothesis 1 can be tested.

2) *Examining research hypothesis 2*
In the second hypothesis, it is hypothesized that community social cohesion based on community social capital are very influential factors on local residents' perceptions towards whether the traditional neighborhoods shall be conserved or be demolished.

To test hypothesis 2, which speculates that community social cohesion constructs have impacts on residents' perception towards neighborhoods conservation, Pearson's correlation coefficient analysis is conducted to examine the correlations of social cohesion constructs and residents' perceptions of neighborhood conservation in each neighborhoods. If data could verify that there exists high correlation coefficient among social cohesion constructs and residents' perceptions of neighborhood conservation,and could identify a certain groups of social cohesion constructs as most influential ones, the hypothesis 2 could be tested.

4.4 Social network analysis under the framework of path dependency

Social network is the one of the three key elements in the field of social capital. "People connect through a series of networks and they tend to share common values with other members of these networks; to the extent that these networks constitute a resource, they can be seen as forming a kind of capital" (Field, 2003, p. 1)

Network is very important to the success of business. Putnam(1993) claimed that in the well-connected societies the whole economic situation is better in the poorly connected societies. Network also contributes to education, health and crime control (Field, 2008; Jütte, 2007).Just as John Field(2008, p. 3) has mentioned,"People's networks should be seen , then , as part of the wider set of relationships and norms that allow people to pursue their goals, and also serve to bind society together."This session aims to illustrate how social networks initiated by different social capital could interact and negotiate the conservation decision making towards a just outcome and process.

4.4.1 Examining research hypothesis 3

In the hypothesis 3, it is hypothesized that in the urban conservation decision-making network,there exist in some patterns of dynamic composition and interactions of different forms of social capital , which are influential to break the lock-in effect in decision making and to negotiate the decision making process towards social justice. In other words, if it can be verified that in the network structure of decision making, several nodes formed by different social capital actively interacted and negotiated, and transformed the conservation process to be socially just in terms of process and outcome,the hypothesis 3 could be tested. To test hypothesis 3,social network analysis under framework of path dependency is employed.

4.4.2 Framework of path dependency

Path dependence has been widely discussed in fields of economic analysis (Liebowitz & Margolis, 1995), institutional change, political change, and technical revolution(Schienstock, 2011).Via the investigation of a development process, path dependence argues that there exists a "lock-inby historical events" until there will emerge a breaking point regarding the returns(Liebowitz & Margolis, 1995, p. 206). Nevertheless, as Schienstock argues, "these developments often are inefficient. On this basis, it is hard to explain how completely new paths evolve". Hence, focus could be shifted to "the creation of new and more preferable paths in the future", and the key role of agency should be highlighted (Schienstock, 2011, p. 63).

In terms of urban conservation planning process, varieties of affecting factors, such as urban policies, conservation approaches, strategies, guidelines, and regulations, together with driving forces form other sectors such as political, economic, and social aspects, as well as interactions between interest groups and organizations, from both interior and exterior, shape and reshape the path of conservation process.

For instance, from physical fabric based preservation, to area based urban conservation, and then to the economic gains oriented urban redevelopment, the path of conservation process has experienced the plan-controlled, cultural intervened, market-oriented decision making pattern. Will social capital and social interactions among the diverse social groups facilitate a transformation of conservation decision making process towards more consensuses and less conflicts?

According to Hoyem (2004), the application of path dependency method could enhance the understandings in historic district planning. In this research, the transformational conservation processes of each case is revealed through a path dependency framework(Magnusson & Ottosson, 2009).The focus is to reveal how social capital initiated interactions could transform the decision-making to be socially inclusive and just.

In the first step, through an inspection of the lock-in effect of the mainstream conservation approach in many Chinese cities, which is profit-led mass demolition and mass relocation, attention is paid to verify whether social capital initiated actions in the transitional stage could function as the driving forces of breaking point or transitional point to a considerably stable path, and reshape, or even create a new path of decision-making pattern.

In the second step, to further examine the dynamic interactions among different relationship nodes, social network analysis is employed to reveal the alternations of decision making network structures through the examination of three stages: the initial stage, transiting stage, and the target stage. The steps is summarized as the following figure (Fig.4-1).

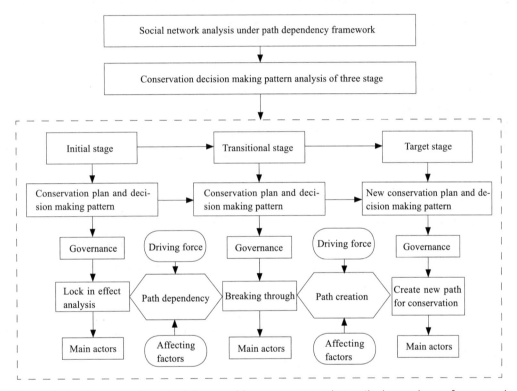

Figure 4-1 Conservation decision-making process under path dependency framework

4.4.3 Social network analysis

Within the micro, neighborhood level, social capital has been measured in terms of cognitive and structural types and the hypothesis related to the neighborhood social capital can be tested through social capital survey. In the meso-level, different forms of social capital, which are manifested in different organizations, formal and informal groups, etc., interact with each other and formulate diverse patterns of cooperation in urban conservation process.

Accordingly, to investigate this interaction and cooperation patterns of meso-level social capital in urban conservation process, specifically in the decision making process, it is also hypothesized (Hypothesis 3) that in the urban conservation decision-making network structure,there exist in some patterns of dynamic composition and interactions of different forms of social capital , which are influential to break the lock-in effect in decision making and to transform the conservation decision making towards a just decision-making process , thus creating new paths for urban conservation practices.

To test this hypothesis, social network analysis under a path dependency framework will be employed. Network analysis contributes to reveal the structure of interaction and cooperation among organizations and institutions(Jütte, 2007). Through social network analysis of a specific condition, the decision making structure in terms of network composition and interaction of this stage could be revealed.

Furthermore, Sandra Franke (2005, p. 17) argues that *"the dynamic component can be summarized as the study of the conditions in which specific networks operate and are mobilized to provide members with access to certain resources: co-operation, support, or capacity building"*. The dynamic interactions among relationship components could be revealed by examining the alternations of decision making network structures through the entire dependency path of urban conservation practices. In the urban conservation decision making process, different nodes of social relationship interact, cooperate and negotiate with each other. When lock-in effect dominated decision making network structure falls, the transitional decision-making network structure rises, hence formulating new development targets and generating trends of decision making network structure based on new targets.

In general, this social network analysis under a path dependency framework will be conducted through the following three steps:

The first step includes the identification and categorization of social relationship nodes of social network structure in each decision making stage.A qualitative social network analysis is conducted in a Meso level. Accordingly in the conservation management process, especially in the decision making process, formal and informal groups, institutions, associations, and organizations are identified as nodes of communication and cooperation, which interact within the Meso level decision making social network and could produce different forms of social capital.

1) *Identification of social relationship nodes*:
The forms of social capital have been discussed widely in terms of their relationship with the community and their different impacts on the community development. In general, interview will identify three forms of social capital in individual/ community level.

Social capital is identified in both the Xi'an Muslim Historic District Protection Project (1997—2002), and the Sajinqiao Redevelopment Project (2005—2007), as well as in every stages of the Tianzi Fang development process. Major organizations and groups, which were involved in the decision-making process of the above projects, are identified as relationship nodes. These nodes communicate or cooperate within or among certain scales, and are categorized by the different relationship ties they could keep within a community.

2) *Categorization of social relationship nodes*
The second step includes the categorization of social relationship nodes of social network structure in each decision making stage. All the relationship nodes will be the categorized according to their relationship pattern with the neighborhood in terms of information and resources exchanging, namely bonding social capital, bridging social capital, and linking social capital(Putnam, Leonardi, & Nanetti 1993). By categorizing the groups, organizations, and institutions according to the three forms of social capital, a network analysis will provide insights to understand the interaction and cooperation between these relationship components, which are initiated by different forms of social capital in the Meso-level conservation decision making at each stage. The detailed introduction of this step will be illustrated in Chapter 7.

3) *Network Analysis —— Social Network structure of decision making*
The third step will examine the characteristics of social network structures developed from the above two steps. According to Nan Lin (2002), "network characteristics", "relations" and "locations" are the key elements when we are determining the access to social capital and the use of social capital in a social relationship network. Social network analysis (Franke, 2005; Putnam, Leonardi, & Nanetti, 1994) in a Meso level will examine the complex interactions and dynamic composition of the identified social capitals in the different stages of conservation practices.

Two parameters will be measured: overall density and degree of centrality. Density is "the percentage of all possible ties that are actually present in a network graph"(Kazmierczak, 2012, p.12). Based on social network theory, a high density of relationship ties positively indicates the implementation of cooperation and innovation (Jütte, 2007; Scott, 2000). Degree centrality means "the number of ties that every node has, or the number of organizations with which each stakeholder exchanged information or collaborated" (Kazmierczak, 2012, p. 12). Degree centrality can indicate the level of a node's communication and cooperation with other nodes. According to literature review, to examine the social network configuration, the examination of network connections should be accompanied by the investigation of network components.

Based on the results of interviews and questionnaires, the research will analyze the social network of each case. This research will use UCINET to carry out the quantitative analysis.[①] Using statistical analysis provided by UCINET, some measurements could be calculated out at the node and network levels; meanwhile the results could be shown in graphics and tables, which might created a visually depict of the social network (Borgatti, Everett, & Freeman, 1999).

Relationship ties of communication and cooperation regarding this project are recorded by the one-way or two-way patterns that are illustrated in the social network of XMHDPP 1997-2002 and SRD 2005—2007 in Hui Fang, Xi'an and in Tianzi Fang's development path.[②] The analysis of data will be discussed in Chapter 7.

4.5 Research framework

4.5.1 Summary of the research methodology

In summary, the data of this research are collected from two channels. One is external sources, which include books, journals, conference papers, governmental reports, maps, news papers and so on. The other is internal source which is gained from fieldwork survey, such as observations, interviews, questionnaires, group discussions.

Once the data is collected, it will be analyzed through qualitative and quantitative methods. For the opinions, depictions and narratives come from interviews and discussions; this research will reveal the perceptions and attitudes and explain the reasons and consequences. For the quantitative data from questionnaires and statistics, the research will use social network analysis tool and software to conduct the analysis and comparisons.

4.5.2 Main steps of the research

The research steps could be outlined as below:

1) *Literature Review*:
Literature review is to collect, and review scholarship on urban conservation and social capital, and to categorize and summarize them in a holistic and systematic way.

---

① As a commercial social network analysis software, UCINET is a comprehensive package and can handle a maximum of 32,767 nodes (Huisman & Van Duijn, 2005), which is adequate for this research.
② For example, when node A approaches to node B to cooperate, consult, report, appeal, or resist, information is sent to node B; if node B seldom cooperates, responds, or consults, the relationship will then be identified as a one-way pattern, from A to B. In contrast, if the communication between two nodes is active and the information is shared between them, accordingly, the relationship will then be identified as a two-way pattern.

2) *Field work*

Field work is to collect information through observation, mapping, and photographs in order to examine the change of functions, social and economic activities, and the effects on the conservation of physical urban fabric and residents' everyday life.

3) *Interview*

The research conducts interviews and group discussions among local residents, tourists, governmental agencies, research institutions involved in urban conservation, developers, stakeholders, etc. in order to gain collective opinions of the forms of social capital and changes of social capital in different stages of conservation decision making process and to understand the perspectives, attitudes and needs of different individuals and groups.

4) *Social capital survey*

Social capital survey and analysis are conducted to assess the quality and forms of stable social relationships among residents of traditional neighborhood which construct community social cohesion; and to examine the impacts of social cohesion on the conservation decision-making process.

5) *Social network analysis:*

The fifth step is to conduct social network analysis in terms of conservation decision-making network structure, which examines the complex interaction and dynamic composition of institutions/associations/organizations in the conservation processby visualizing social network.

6) *Discussion of findings*

Based on the findings, this research will discuss the implications of social capital for urban conservation decision making, and will propose a conceptual urban conservation framework to integrate social capital approach with current urban conservation planning in China.

4.5.3 Research framework

In short, the framework of this research could be summarized as the following figure (Figure 4-2)

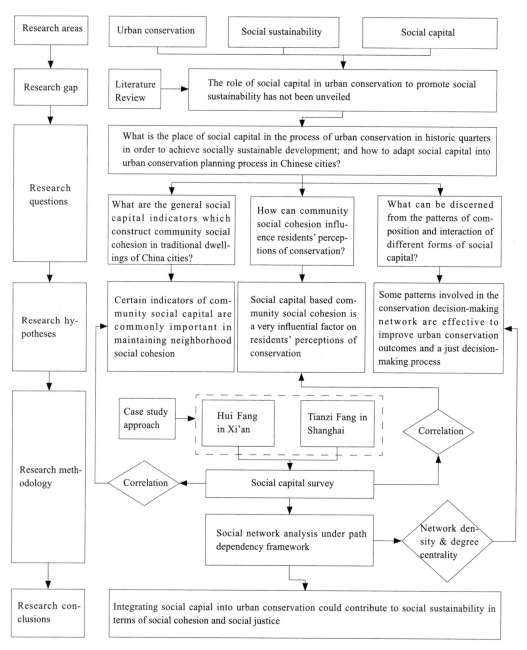

Figure 4-2 Research framework

Chapter 5 Case Study of Hui Fang in Xi'an

Xi'an is a historic city located in northwestern China. The current townscape originated from Xi'an Fu during the Ming Dynasty in 14th century AD. The townscape consists of circumvallation with a moat, the Drum tower, the Bell tower, and several avenues. Located in the city center, Xi'an Muslim District covers 1.3 square kilometers and accommodates over 60,000 residents, among which approximately 35,000 are Muslims. The history of this area as a district of Muslim concentration can be traced back to Song Dynasty (960—1279) (Bi 2004; Chen 2005). Within this district, there are 11 mosques situated in 13 neighborhoods of traditional courtyard housing. A unique social-spatial unit "SI-FANG" (mosque-neighborhood) was formed gradually as Muslim people settled around these mosques, running small-scale business alongside neighborhood market streets.

5.1 Introduction of Ethnic Hui

5.1.1 The origins of Ethnic Hui in China

In Tang and Song dynasty, the central government at that time adopted an open policy in order to encourage trade with overseas countries. As a result, quite a lot foreigners immigrated to China, and were called as "Fan Ke(藩客)" in general in the beginning (Ding 2005).Among their descendants,the identity of Hui-Hui was recognized as immigrated Muslims during Yuan Dynasty (1271—1368) because their sense of ethnic origin became stronger and stronger after they lived, worked and fought together for a long period. Finally, during Ming Dynasty, the identity of Ethnic Hui was well recognized (Ibid.). According to Ding (Ibid.), the history of "Hui-Hui" to be officially recognized as the Ethnic Hui can be traced back formerly to Qian Long Period (1735—1795). Ever since then, Ethnic Hui is officially identified as an ethnic minority.

In China, Ethnic Hui has as trong identity, which can be shown by the low rate of intermarriage with other ethnic groups (Li 2004). The strong ethnic identity might be attributed to their Islamic faith (Editing-Group 2009). However, Ethnic Hui in China is also distinguished with other Muslim ethnic group in the world due to the influence of Han culture or Confucian culture. Firstly, the ethnic origins of Ethnic Hui were diverse and Chinese became their common language after Tang Dynasty with the purpose to adapt to Chinese condition of the day(Li 1990). Using Chinese became a special identity of Ethnic Hui which made it different from other ethnic groups and helped its culture development as well (Ding 2005). With extracting the essence from Chinese culture, Hui people's identity was gradually influenced. In Ming Dynasty, the development of Islamic religion was impeded because central government adopted assimilation policy to minorities. According to Ding (2005), Islamic religious texts were translated into Chinese and some Confucian philosophies were infused along the interpretation and translation. For example, Ding (Ibid.) discussed the special phenomena of Ethnic Hui in Ming Dynasty, which is the transformation from unitary loyalty (be loyal to Allah) to dual loyalty (be loyal to Allah and emperors) in order to adapt to Chinese social pattern and the imperial ideology in ancient China, which is an important essence of Confucian culture.

5.1.2 The origin of Ethnic Hui in Xi'an

The origin of Ethnic Hui in Xi'an can be traced back to Tang Dynasty (Hao 2009). Along with the Silk Road business activities which started from the east end city——Tang Chang An, Islamic businessmen and religionist from Arabian countries and Persia came to Tang Chang An (Han 2003). According to Ma (1981), Ethnic Hui in Xi'an are constituted by four big branches. The first branch is Fan-Ke (蕃客), which is the general name for foreign visitors, businessmen, religionist, and musicians. The second branch is Ya-Shang (牙商), mostly Shiite Muslim refugees, who settled down in northern China and started business between China and other countries, the third branch is the reinforcement from the Arab Empire to quell the An Shi Rebellion (安史之乱) in Tang Dynasty. The reinforcement was allowed to reside in Chang An. The last branch is the small nine tribes from Zhao Wu area (昭武九姓胡). During the eighth and ninth century, these nine tribes in Zhao Wu area were all conquered by the Arab Empire, and those who resided in Chang An were converted to Islam (Wang 1997).

The four branches above settled down and became the main ancestors of the Ethnic Hui in current Xi'an city. Their descendants integrated gradually into the custom of Han society along with history. Their lifestyle and mode of behavior significantly changed along the process of Hanification (汉化).

At the turning of Yuan and Ming Dynasty, Ethnic Hui in Xi'an became an ethnic minority group and Xi'an became a centre for the education and academy of Islam in China in 16th century (Li 2010). In Qing Dynasty, the development of Ethnic Hui was slowed down. During the Anti-Japanese War in 1940s quite a lot of Hui people in Henan, Shandong, and Hebei provinces moved to Xi'an in order to escape from the warfare and Japanese bullying, and the population of Ethnic Hui increased again (Ibid.).

According to Li (Ibid.), there were 54,981 Hui people in Shanxi Province in 1953 and about 20,000 Hui people lived in Xi'an city; in 2000, there were 64,261 Hui people in Xi'an city and about half of them inhabit in Lianhu District.

5.2 The evolution of Hui Fang before1980s

Influenced by their religion and ethnic identity, traditional Ethnic Hui dwell collectively in nearby area, which popularly formed the tradition quarters of Ethnic Hui, such as Hui Street, Hui Hutong, and Hui Village. After years of evolution and fusion, these quarters have evolved into a typical form of social organization among traditional communities of Ethnic Hui in China——Si-Fang（寺坊）which are mix-used, combining residential, commercial and religious functions.

The evolution of Si-Fang structure in Xi'an for Ethnic Hui can be categorized into to five stages. It was originated with "Fan-Fang" in Tang-Song Dynasty as the settlements of Ethnic-Hui's ascendants, was transformed into a typical Mosque–centered settlement pattern along with the establishment and development of Ethnic Hui society in Ming-Qing Dynasty, and was eventually evolved to "Twelve-Mosque"structure in 1980s.

5.2.1 Fan-fang in Tang Dynasty

Tang and Song Dynasty (618-1279) could be seen as the original stage. As described earlier, due to easy and comfortable foreign policy and openness of society, considerable Fan-Ke（蕃客）—— businessmen and visitors from northern and western countries came to China. Fan-Ke dwelled closely in big cities such as the Capital city ——Tang Chang'an and port city Quanzhou. Their agglomerations were planned as Fan-Fang（蕃坊）, which can be identified as the early form of Si-Fang for Ethnic Hui. In Tang Dynasty, government has established an administrative system for Fan-Fang. A leader was nominated as Fan-Zhang（蕃长）and "Fan-Zhang-Si"（蕃长司）and established as the administrative organization of Fan-Fang (Fang 1996).

Built based on the city of Sui Daxing, Tang Chang'an is a well-planned city based on the "Shi-Fang" (Markets-Residences) pattern. Based on the archeological findings and documents, the city is surrounded by city wall, which is 8652 meters from south to north, 9721 meters wide from east to west, covering an area of 84.1 square km (Dong 2004). The northern city wall has two gates, while the rest all have three gates on each side. There are three main avenues and three main streets within the city wall area. A further subdivision has divided the city into grids by nine avenues and twelve streets. Except the large scale royal buildings such as the Imperial City and the Palace, the rest grids are planned as 110 Fangs (residences) and 2 Shis (markets) according to the Shi-Fang pattern (Fu 2001).

In Tang Dynasty, Shi-Fang was adopted as a basic planning principle. Fang as a typical spatial term was generated with specific historical and cultural backgrounds. According to Tang Liang Jing Cheng Fang Kao (《唐两京城坊考》), the 110 Fangs within Chang'an City were under enclosed management, which means that business activities were limited strictly within Shi——market area, namely the East Market (东市) and the West Market (西市), while the residence are arranged within Fang——residence area (He 1980). According to Wu (1985), the East Market, which is adjacent to the Royal Palace and governmental offices, mainly sever the noble class and bureaucrats. As an international commercial quarter for the citizens, the West Market is with close access to the west gates of the city, through which the troops of foreign businessmen came to the city of Tang Chang'an (Ibid.). According to the record of ancient books Tai Ping Guang Ji (《太平广记》) and Zi Zhi Tong Jian (《资治通鉴》), the shops of foreign businessmen from countries like Arab and Persia were mostly located near the East Shi (Market) and West Shi (Market) area, especially the West Shi area (Huang 2010).

In the early stage of Tang Dynasty, a large number of foreign businessmen from countries like Arab and Persia came to Tang Chang'an, seeking business opportunities. It is recorded that many businessmen settled down and refused to return. Their collective dwelling areas were called Fan-Fang (蕃坊). According to the Ge (1998), the temples of Zoroastrian, which is the national religion of Persia, were built inside many Fangs, such as Bu-Zheng Fang (布政坊), Li-Quan Fang (礼泉坊), and Chong-Hua Fang (崇化坊), etc. As temples normally have a considerable group of believers; it can be assumed that a considerable number of foreigners dwelled in these areas (Figure 5-1) (Huang 2010). As these Fan-Fangs are adjacent to the West Market, it also can be assumed that good access to market was the priority for foreign businessmen. Huang (Ibid.) developed a spatial pattern of Fan-Fang, which is Shi/market centered dwelling pattern.

Meanwhile, it is also noticeable that Fan-Fang in Tang Dynasty embodied cohabitation of multi-ethnic groups, which led to not just economic exchange, but also interactions among them towards cultural blending and hybrid dwelling (Ibid.).

Due to the rebellions and wars in last stage of Tang Dynasty, Tang Chang'an decayed gradually, and the Capital has been relocated to the city of Luoyang number of times. In 904 AD, Han Jian, the governor of Chang'an rebuilt a largely downsized city as a defensive military city, called Han-Jian New Town (Wu 1999). The city was only rebuilt on the site of the previous Imperial City area. As residents gradually moved into this new town, the ascendants of Ethnic Hui gathered near the North West corner of the city, and formulated the settlements of Ethnic Hui (Ma 1981).

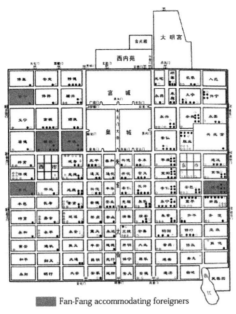

Figure 5-1　Fan-Fang in Tang Dynasty
Source: Huang, 2010, p.28.

Nevertheless it is in Song Dynasty (960-1279) that the pattern of Hui Fang was established in the city of Song Chang'an. During Song Dynasty, the settlements pattern of the ancestors of Ethnic Hui developed gradually based on the pattern of Fan-Fang. Hybrid dwelling of multi-ethnic groups continued to be a feature, while in the same time, with the economic development, the rigid separation of market and Fang collapsed. The walls of Fang were broken for shops, and the commercial space extended from market areas to the street shops along the edges of Fang (Li 2007; Jiang 2012).The further openness of Fan-Fangs was initiated by the extension of the market space to the streets on the edges, and the space pattern of Fan-Fang as a close residential unit transformed gradually into the mix-used Street-Alley space, which identified by commercial streets and residential alleys, could accommodate diverse functions such as commercial, residential, administrative uses, etc. (Li 2007). The institutional shift from "Li-Fang" in Tang Dynasty toward "Street-Alley" and Song Dynasty contributed to the stabilization and agglomeration of the ascendants of Ethnic Hui's community, which has established the early form of Hui-Fang by adding the elements of commercial streets to the spatial structures of Fan-Fang and extending the openness to the spatial characters of Fan-Fang (Huang 2010).

5.2.2　Si-Fang in Yuan Dynasty

The second stage is during the Yuan Dynasty (1271—1368). Aftermath the success of Western Expedition of Mongolia in the first half of 13 century, a great number of Muslim soldiers and craftsmen from Persian, Arab and Middle East countries were allocated to the Mongolia Army in order to support the unification war within China. After China was united under Yuan (1276) Dynasty, these "outsiders" are relocated to the villages and communities in Northwest China. Following the Fan-Zhang-Si (蕃长司) in Tang and Song Dynasty, The government of Yuan Dynasty established Bureau of Hadji for Hui-Hui's autonomy, which took in charge of the Muslim community affairs (Qiu 2001).

Under an autonomy administrative system, religious teaching together with Mosques played an increasingly important role in the Muslim community, which laid a strong foundation for the Si-Fang (Mosque-Residence) structure. In the early Yuan Dynasty, the population of Muslim in Xi'an increased because the central government adopted friendly policy to foreign racial groups in order to suppress the development of ethnic Han (Huang 2010). Meanwhile, the religion of Islam was recognized as main religions as Buddhism, Taoism, and Christianity; accordingly, quite a lot of mosques were set up for worship and other religious activities (Ibid.). Assumingly, Hui people were willing to reside nearby the mosques instead of markets with the intention of religiosity and convenience. In summary, Hui Communities gradually transformed from Market-centred from to Mosque-centred form during Yuan Dynasty, which is the embryonic form of "Si-Fang" structure after Yuan Dynasty.

5.2.3　Hui-fang in Ming-Qing Dynasty

The third stage is during Ming and Qing Dynasty (1368-1912). To cultivate support from minorities, in many areas, Mosques were built by imperial order or from local funding. Meanwhile to enhance the power centralization, an entire system was established from the central government to local governments (Guo 1983). Different from "Hadji" of Yuan Dynasty which was autonomy, Ming Dynasty employed "Three-Zhang-Jiao (三掌教)" system as the religious manage system, which means that the Hui communities are mutually led by three leaders, namely *Imam* (伊玛目), *Khatib* (海推布),and *Mu'adhdhin* (穆安津), who cooperated in religious affairs of the Hui community (Wang 2002). Regarding their responsibilities, Imam is the religious leader, who presided over religious rituals and activities; Khatib was in charge of religious propaganda and education; and Mu'adhdhin is the caller of worship and congregation (Ibid.). In the second half of 14th century, along with the formation of Hui-Hui as an ethnicity, *Imam* was officially recognized as the leader of the Ethnic Hui and became hereditary."Three-Zhang-Jiao"gradually transferred to the "Iman-Zhang-Jiao", which has played an important role for the growth of Muslim in China and the establishment of Si-Fang unit in traditional Ethnic Hui community.

In late Ming Dynasty, the Islamic education in Mosques(经堂教育) prevailed in Guan Zhong (关中) district that is a historical region of China corresponding to the lower valley of the Wei River (Wang 1993; Xue, Yao et al. 2000), and Mosques played an important role in the formation of Si-Fang's spatial pattern. According to Li (2010),Xi'an at that time was the center of Islamic culture and Islamic education. Great quantities of Mosques, which bear the function of Islamic education, were built among Ethnic Hui's neighborhoods.It can be identified that Mosques also called Islamic temples among Ethnic Hui people evolved to be the centers of Ethnic Hui's community.

According to Ma (2012) there existed two forms of Si-Fang, namely Si-Fang in cities and in countryside. Si-Fangs in countryside have shown clear boundaries. Normally a Si-Fang in countryside was centered by a mosque and was surrounded by several villages of Ethnic Hui.

For the Si-Fangs in cities, the spatial characters remained as mosque centered, while the close physical boundaries of villages were replaced by the open streets and alleys system, with a certain amount of nearby residents as stable believers. There was no clear boundaries as the spatial characters of Fang have evolved from the close pattern of Li-Fang to the open pattern of Street-Alley. In addition, though residents of Ethnic Hui normally belong to a certain mosque, they may join Mosques located outside of their neighborhoods, since there gradually emerged several branches in Islamic mosques since Ming Dynasty (Wang and Ma 1982). With the enhancement of Islamic education, Si-Fang in cities has become a social spatial concept, which functioned as not only a residential neighborhood unit of street and alleys to accommodate Ethnic Hui , but also as a social unit to encounter their rich social life, including Islamic culture education, religious activities, trade businessand so on .

Different from Fan-Fang in Tang Dynasty which was autonomous, Si-Fang in Ming and Qing Dynasties was managed under the local administrative system, and had no independent administration power after the separation of religion and administration in late Yuan Dynasty (Wang 2002). Si-Fang instead took in charge of Religious affairs and those everyday life activities related to Ethnic Hui's religious life.

In Yuan Dynasty, a basic social unit——SHE (社) was organized under Fang to maintain the social order . In rural areas, 50 households were identified as a SHE; while in cities, SHE did not have restricted household numbers. The head of SHE will help maintain social order of neighborhoods.

In terms of Mosques and neighborhood management, the existed another system called "She-Tou (社头)" or "Xue-Dong (学董)". Xue-Dong can be traced back to the rising of Islamic Jing-Tang Education. When Islamic Jing-Tang education was given priority in Mid-Ming Dynasty, believers of an Islamic neighborhood elected the patron from affluent families to be Xue-Dong to support the religious teachers and students and to promote Islamic Jing-Tng Education. According to Ma(2012), She-Tou normally was elected and took charge of the management of mosques.

It was recorded that in the city, there were 7 Mosques within the inner city area (Ma 2008; Ma and Ma 2010; Ma 2012). According to Ma (2012), these 7 Mosques refer to Hua Jue Alley Mosque（化觉巷清真大寺）, Dapiyuan Mosque（大皮院清真寺）, Xiaopiyuan Mosque（小皮院清真寺）, Daxuxi Alley Mosque（大学习巷清真寺）, Beiguangji Street Mosque（北广济街清真寺）, Qingzhenyingli Mosque（清真营里寺）, and Sajinqiao Ancient Mosque（洒金桥清真古寺）, which were all located within the current Ethnic Hui District. It was well accepted that during Mid-Qing Dynasty, there existed a well-known pattern called "Seven Mosques and Thirteen Fang", which means that these seven mosques were surround by thirteen neighborhoods characterized in an obvious Street——Alley pattern. According to Yang (2006) ,the thirteen Fangs include: Xuan-Ping Fang, San-Xiang Fang, Tie-Lu Fang, Guang-Ji Fang, An-Ding Fang, Bao-Ning Fang, Xin-Xing Fang, Qian-Suo Fang ,Tie-Lu Fang, Xiang-Mi-Yuan Fang, Nan-Shun Fang, Gong-Yuan Fang, and You-Suo Fang. However, according to Ma (2008) and, with the emergence of historical events such as wars, Seven Mosques remained stable, while the numbers of Fangs can't be identified. Seven Mosques and Thirteen Fang might be a general description or a perception of the scale and layout of Ethnic Hui community in Xi'an, while the stable Si-Fang pattern has lasted for hundreds of years. Si-Fang as a unique social-spatial unit has been deeply imbedded in Ethnic Hui's perception to their communities. Till now it still plays an important role in the social life and everyday activities among Ethnic Hui residents.

In Qing Dynasty, Islamic Region experienced schism in Northern-West regions such as Gan Su, Qing Hai and Xin Jiang, where a new Islamic religious sect called Sufism prevailed. According to Ma (2010), in Qing Dynasty Tongzhi Period, many Hui people outside Xi'an City migrated towards West along the Silk Road due to certain political reasons, which prevented the introduction of prevailing new branches into Xi'an. Fortunately the Ethnic Hui community in the inner city of Xi'an survived, and Qadim with Three Zhang Jiao Teaching System remained as dominant Islamic religious sect in the seven mosques, with the hereditary system for Imam was gradually replaced by appointment system for Ahong (Ma and Ma 2010). The general concept of Hui Fang as Seven Mosques and Thirteen Fang, especially the Seven-Mosques pattern sustained in Tongzhi period and succeeding eras in Qing Dynasty.

5.2.4 The structural change of Si-Fang in the period between 1912 and 1949

In this period, a Nine-Mosques pattern emerged in Hui Fang area as the result of Ikhwan[①] ref-

---

① Also known as Muslim Brotherhood, more information please refer to http://en.wikipedia.org/wiki/Ikhwan.

ormation movement in Xi'an. According to the record of Ma (2010), in this period, Ikhwan faction of Islamic Religion was introduced to Xi'an from Gansu province. In Da-Xue-Xi Alley and Xiao-Xue-Xi Alley areas, two Qadim mosques, called Da-Xue-Xi Alley Mosque and Xiao-Xue-Xi Alley Ying-Li Mosque, were converted to Ikhwan Mosques. As a result, though remaining as residents of these two streets, some faithful Qadim followers left these two mosques and built another Qadim Mosque, called the Middle Mosque（清真中寺）in Xiao-Xue-Xi Alley in 1919. In Sa-Jin-Qiao area, due to the broadcast of Ikhwan, some previous Qadim followers of the Old Mosque（清真古寺）left and built the West Mosque（清真西寺）in 1920 (Ibid.).

With the development of Ikhwan Religious sect in Xi'an in this period, the social and spatial structure of Si-Fang as a spatial unit was largely affected. In terms of the relationship between Fang and Mosque, the previous direct link between residential areas and their respective nearby Qadim Mosques was broken. Those who live in the same streets could be attached to different Mosques. In terms of demographic structure, as members of families believed in different Islamic Religious sect, big families were divided into several households, and some moved to follow their own Mosques.

5.2.5 Hui-Fang's Twelve-Mosque structure in the period between 1950s and 1980s

This period saw the closure of most of the Mosques due to the disturbance of the Cultural Revolution Movement, the sub-division by the introduction of Salafiyya[①] Islamic fraction to Hui Fang, and the rehabilitation of Hui Fang after the Cultural Revolution Movement.During the Cultural Revolution Movement, except the Great Mosques, all the other Mosques in Hui Fang were closed for religious activities. And most of the Mosques were changed to facilitate factories or other uses during this period. It was in 1982, the religious education was rehabilitated and religious activities came back to Mosques. The previous Nine-Mosque pattern was recovered.

According to Ma and Ma (2010), Salafiyya was firstly introduced to Hui Fang in 1950s. But the introduction of new religious sect was stopped by the Cultural Revolution Movement. After the Cultural Revolution Movement, two Salafiyya Mosques were established, which are Hong-Pu-Jie Mosques and Xi Cang Mosque. Some of the Hui people converted to Salafiyya and left their previous Mosques. In 1990s, another Qadim Mosque, called Lv-Shan Mosque was established in Xiang Mi Yuan to accommodate the migrant Hui people.

With the emergence of new Mosques, the previous Nine-Mosques pattern shifted to Twelve-

---

① Also known as Salafi movement, more information please refer to http://en.wikipedia.org/w/index.php?title=Salafi_movement&redirect=no.

Mosque in 1980s (Ma 2011). Residents living in the same neighborhood——Fang may not attend the Mosque adjacent to the identical Fang.

Rather than representing the previous concept of Jiao-Fang, which has both clear boundaries in physical and social aspects,the Twelve-Mosques pattern has established the basic structure of modern Hui Fang (Figure 5-2). It represents the clear neighborhood and Mosques boundary in physical level, which although is not representing clearly in social level, has significant influence in the enhancement of identity and the construct of the sense of belongings through followers' everyday religious practices and festival gathering (Figure 5-3, 5-4, and 5-5).

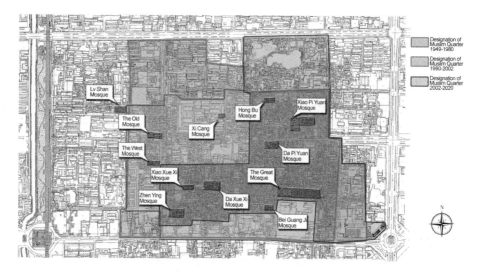

Figure 5-2 Twelve-Mosques pattern in Hui Fang

Figure 5-3 Prayer service in the Great Mosque

Figure 5-4   After Friday prayer service in Dapiyuan Mosque

Figure 5-5   Celebrating the Feast of the Sacrifice-Corban Festival in the courtyard of the Great Mosque

5.3   Urban fabric of Hui Fang community in Xi'an

Current Hui Fang covers 1.3 square kilometers and accommodates over 60,000 residents, among which around 35,000 are Hui people (Han, Wang et al. 2015).Since Tang Dynasty, Chang'an City (current Xi'an City) was planned and constructed with the regular grid street system. Ever since, this special urban fabrichas remained and become a model for other ancient Chinese and Japanese cities. In Xi'an, the urban structure of Hui Fang is also characterized by such fabric, which can be seen in Figure 5-2. According to Cao (2005), a spatial structure with strong order and hierarchy can be discovered in Hui Fang, within which the intercrossing linear streets and alleys can be seen as framework, courtyard houses along the streets can be seen as cells, and individual buildings shape courtyards through certain patterns. This spatial structure is the foundation of the unique townscape of traditional historic areas.

In Figure 5-6, it can be seen that the principal axes of courtyard houses are mostly perpendicular to the main street so that each courtyard houses can equally acquire enough space for shops along the streets. According to Cao (Ibid.), most of the courtyard houses in Hui Fang have one side facing to the street, which is normally about 10 meters wide and divided by 3 bays. Besides, the first floors of the buildings along streets are usually used as retail space, and their upper floors are residential space, while the inner space is usually for producing and processing (Ibid.). In addition, courtyard houses are normally composed by several small narrow-long courtyards as it is shown in Figure 5-7.

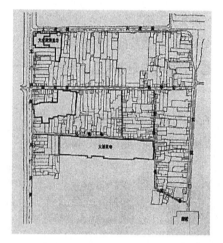

Figure 5-6   Urban fabric in Hui Fang
Source:(Cao 2005), p.9.

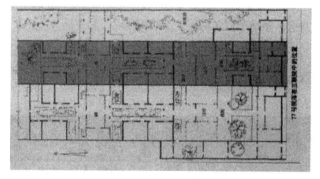

Figure 5-7   Typical courtyard house in Hui Fang
Source:(Cao 2005), p.9.

5.4 The street life and social relationship in Hui Fang

In this case study, in order to reveal the charateristic of street life and social relationship in Hui Fang, three methods were adopted, including fieldwork, obseravation and survey.

5.4.1 Streets analysis

According to an unpublished study by Lv (1997), the street in Muslim quarter could be divided into three categories. The first category is main streets which serve as public space, such as Sajinqiao Street. The second category is defined as neighborhood streets, which serves as semi-public space for local people, including Guangji Street, Xiyangshi Street and Miaohou Street. Based on observation, it is found that most of commercial activities in Muslim Quarter might be flourishing in main streets and neighborhood streets. In addition to these two, the third category includes other small and winding alleys and cul-de-sacs. Most of these alleys serve as links and connect dwellings and other streets; therefore they also can be described as semi-private space. It was also recommend that this special urban fabric and street structures should be preserved.

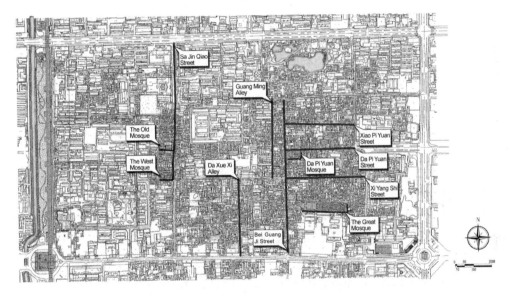

Figure 5-8 Selected streets and mosques for observation

In this research, as shown in Figure 5-8, seven streets are chosen as case studies to analyze the street functions, including Sajinqiao Street which is main streets, Daxuexi Alley, Guangming Alley, Bei Guangji Street, Dapiyuan Street, Xiaopiyuan Street and Xiyangshi Street which could be considered as neighborhood streets. Meanwhile, based on field observations, Xiyangshi Street is selected as special case to discuss the activities of different populations. In the following section, these cases are introduced in details.

1) *Sajinqiao Street*

With the length of 420 meter, Sajinqiao Street is a main street in Muslim Quarter. Together with its extension Da Mai Shi Street, it runs through the north and south of the Muslim quarter. Interestingly, the name of the street comes from a legend recorded in "Tang Shu". According to this legend, the great emperor "Tang Xuan Zong" in Tang Dynasty once spread gold on a bridge in this street to his officials. Therefore the meaning of the street name is to record this moment and literally means the bridge where somebody had spread gold on. Nowadays, it is very important street for the Muslims because there are two important mosques located in this street, namely The West Mosque and The Old Mosque.

Initially, the commercial activities in the street are very prosperous due to its important location. Lots of Muslims opened their businesses in Sajinqiao Street, mainly groceries and catering restaurants. Besides, there were also quite a lot residential buildings located on both sides of the streets. However, in 2004, with the purpose to improve the transportation connectivity of this area, Xi'an municipal government tried to expand this street and demolish the buildings on both sides. Nonetheless, due to the collective actions of conservation from local residents, this project could not be carried out successfully and has to be suspended until now. In Figure 5-9, it can be seen that a number of buildings on the street have been demolished. In addition, with the concern of being demolished, many shops, particularly catering shops remain closed for a long time, albeit they are not demolished. Accordingly, the commercial activities are strongly affected by this event. As such, this street is suffering from the decay caused by physical deterioration.

2) *Xiyangshi Street*

Xiyangshi Street (Figure 5-10) is located at northwest of Drum Tower, and connects Bei Yuan Men Street and Bei Guangji Street. The history of this street could be traced back to Yuan Dynasty. At that time, this street was named by Yang (Sheep——羊) Street because most of transactions of sheep and lambs happened here. But in Qing Dynasty, another street located in the east of city was also famous due to transactions of sheep and named as Dong Yang Shi Street (East Sheep Street). Accordingly, this street finally was called as Xiyangshi Street (West Sheep Street) in order to distinguish the two "sheep streets" until now.

This street serves as a neighborhood street and its length is about 244 meters. Nowadays, this street is famous due to its delicious Muslim food and attracts thousands of tourists every day. Figure 5-10 demonstrates the function layout of this street, in which it can be seen that most of street space was occupied by catering restaurants, snack food shops and clothing stores, which are believed as business targeting at tourists. However, there still are some Muslims live in upper floors. Therefore, contractions between restaurant owners and inhabitants do exist. Besides, lots of venders also run their business on the street to pursue profit, which results in serious traffic congestion in rush hours.

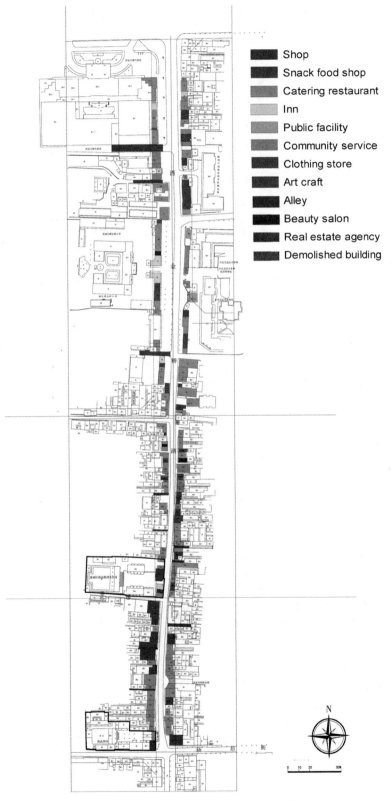

Figure 5-9　Sajinqiao Street is now facing decay

Although, the most of buildings in Xiyangshi Street are occupied by tourism businesses, there are still some public facilities which serve local inhabitants (Figure 5-11). For instance, the electricity power service, public toilet and garbage collection can be found which mainly provide basic needs for residents. Meanwhile, in order to meet the medical and cultural needs, a pharmacy, a public bath, a wedding planner studio, and a photography workshop can also be found in the street. In addition, there are also some shops served as supplier for other restaurants, such as a stove store, two meat shops and one Muslim bread shop.

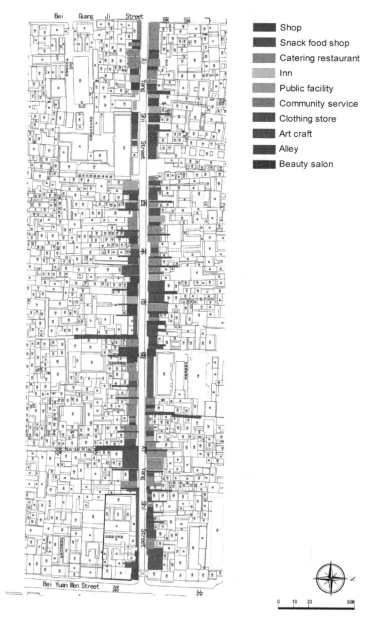

Figure 5-10　Mix-used functional layout along Xiyangshi Street

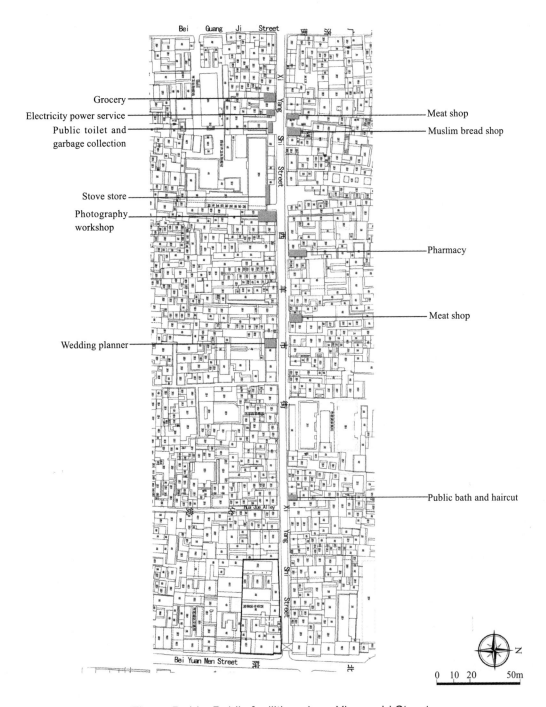

Figure 5-11 Public facilities along Xiyangshi Street

3) *Daxuexi Alley*

Daxuexi Alley might also be considered as a neighborhood street, whose length is about 390 meters. It links The West Avenue and Miaohou Street. Since Tang Dynasty, this street had become the place for Muslims to gather around. According to historical record, Muslim im-

113

migrants from Persian Empire, Caliphate, and other ancient Arab countries lived here and the Tang government established agencies and institutions to teach them about Chinese culture and etiquette. Therefore, together with another alley, Daxuexi Alley was considered as a point for Muslim immigrants to settle down and learn how to communicate with Chinese people. Therefore, in order to recall such history, the name of Daxuexi Alley literally means a large alley for education.

Nowadays, the function to educate Muslim immigrants do not exist anymore. Similar with other streets and alleys in Muslim quarter, Daxuexi Alley is a commercial street for traditional Muslim food and handicrafts (Figure 5-12). Interestingly, there are also quite a lot of beauty salons could be found in Daxuexi Alley.

In Daxuexi Alley, most of basic needs for everyday life could be met. For instance, a fruit shop, an aquatic product shop, a mobile phone store, an electrical maintenance service, and two hardware stores can be found in the street (Figure 5-13). Meanwhile, there is a pharmacy and a clinic in order to meet residents' medical requirement. In addition, Daxuexi Mosque is quite an important religious place for local Muslims. Therefore, there is a shop near the mosque which is selling religious goods. Moreover near Daxuexi Alley, Xi'an Chenghuang Temple is located. Conventionally, traditional drama will be performed in this temple on special occasions, and some local residents are involved in these events and employed as actors and musicians. To facilitate these religious events one costume shop and two musical instrument stores are located in this street.

4) *Guangming Alley*
With a 406-meters-length, Guangming Alley is a second level street according to Lv's classification. However, compared with other streets and alleys, Guangming Alley is relatively narrow and quiet. Since Ming Dynasty, It was named as Guoqianshi Alley to memorize a special education official, whose family name is "Guo" and official position is "Qian Shi". During culture revolution in 1970s, together with most of the streets, the name was changed to Guangming Alley, which literately means bright or promising. After culture revolution, although most streets' name were changed back to its former name but, Guangming Alley still keeps this name until now.

Nowadays, Guangming Alley is not as popular as other streets in terms of commercial activities. Only a few catering restaurants could be found in south end. Most of buildings on the street are occupied by residents, community services, and public facilities (Figure 5-14 and 5-16).

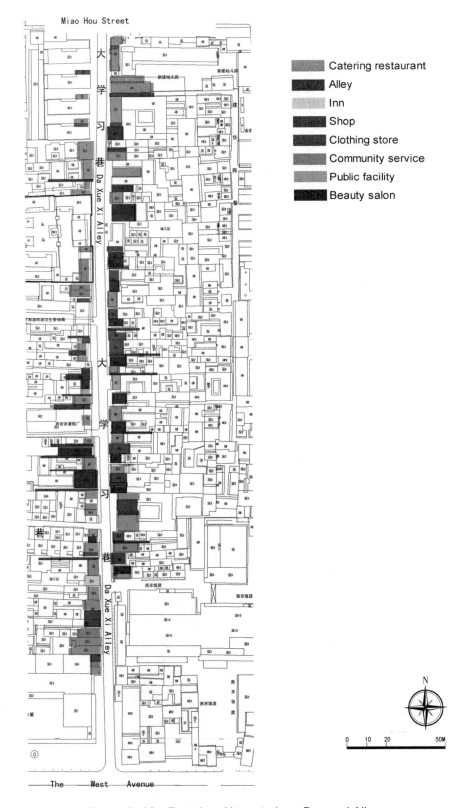

Figure 5-12　Functional layout along Daxuexi Alley

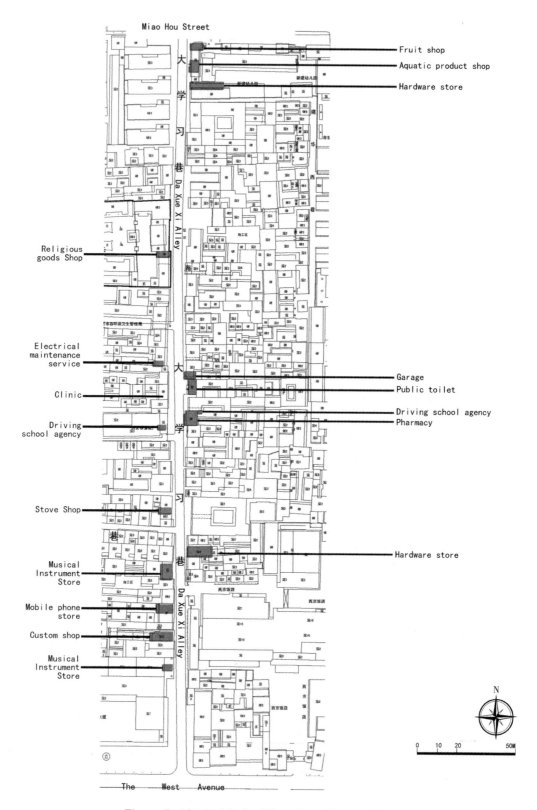

Figure 5-13 Public facilities along Daxuexi Street

Compare to other streets, the main function of Guangming Street could be defined as residential street. In the north part, most of non-residential buildings are occupied by community service and public facilities, including not only groceries, laundries, but also electrical maintenance services Locksmith Shop, and pharmacies, sweets, tailors, photography workshop and gym. Besides, some special facilities, such as building material store and interior design studio also can be found in this alley. In addition, in Daxuexi Alley, there are also a traditional gin house and a steamed bread store, which is relatively rare in modern cities.

5) *Bei Guangji Street*
Bei Guangji Street is the longest street compared with other cases in this research; the whole length is about 800 meters. It connects The West Avenue at the south and Hong Bu Street at the north. It was defined by Lv as a neighborhood street although it is more important than normal secondary streets in terms of transportation and commercialization. In Tang Dynasty, this street once was the widest street in the capital of Chang-an (Current Xi'an). At that time, the street name was "Cheng Tian Street" and also known as "the street to heaven". In Song Dynasty, in order to upgrade water supply system, a canal was constructed and passed through here, which was named as "Guangji Canal". Therefore, the street name was changed as "Guangji Street" accordingly. Divided by The West Avenue, the north part of the street in Muslim quarter now is known as "Bei (North) Guangji Street".

Currently, there are three mosques in Bei Guangji Street, namely The New Mosque in the north end, Dapiyuan Mosque in the middle, and Bei Guangji Mosque in the south part. Furthermore, there are catering restaurants and snack food shops located in Bei Guangji Street, especially in the south part (Figure 5-14 and Figure 5-15). In the north part, clothing stores and art and craft shops can be found, which mainly serve tourists.

Meanwhile, quite a lot community services and public facilities also locate in the south part in order to improve local inhabitants' life quality. Figure 5-16 and Figure 5-17 show all the public facilities and community services in the north part of Bei Guangji Street, which are relatively comprehensive than other streets (Figure 5-16), especially in terms of community services. Regarding basic needs for everyday life, mobile phone stores, groceries, laundries, clinics, pharmacies, electrical maintenance services, and vegetable markets are all accessible. Meanwhile, other uncommon services, such as building material stores, garages, coal shops and grain shops also can be found. Besides, there are also some traditional handicrafts, including a gin house, a tailor's shop, a smithy and a special shop offering service to process golden and silver jewelry. In addition, some raw materials for traditional Muslim food also could be purchased here, such as homemade noodle and sesame oil. For the purpose to improve Muslims' culture and entertainment life, bookstores, matchmaking agencies and travel agencies are also located in north part of Bei Guangji Street. It might be possible that these community services are not only for inhabitants in the street but also for others in Muslim quarter. Therefore, it might be assumed that in Muslims' minds, this street is considered as a community center.

Figure 5-14 Functional layout along Guangming Alley and Beiguangji Street (northern part)

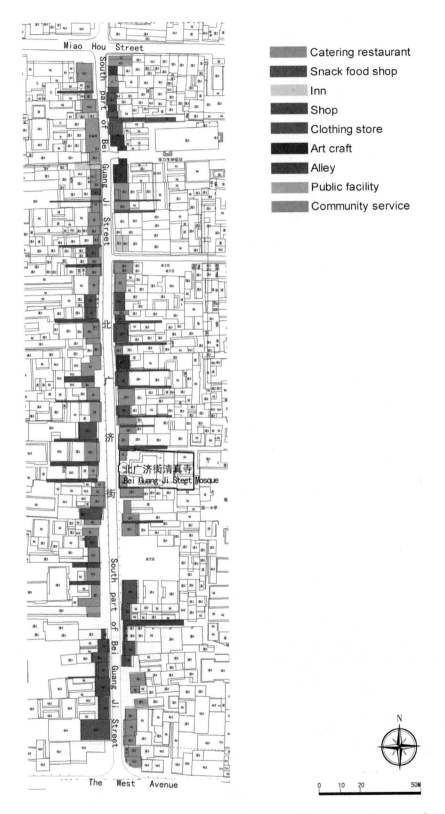

Figure 5-15　Functional layout along Beiguangji Street (southern part)

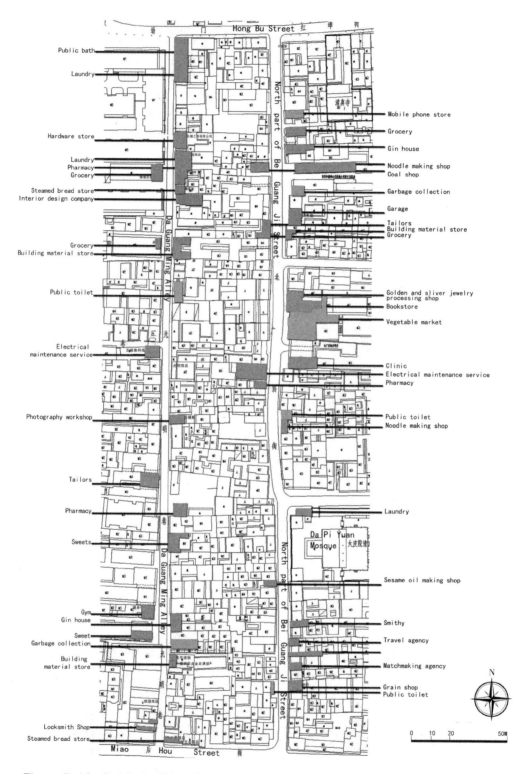

Figure 5-16 Public facilities along Guangming Alley and Beiguangji Street (northern part)

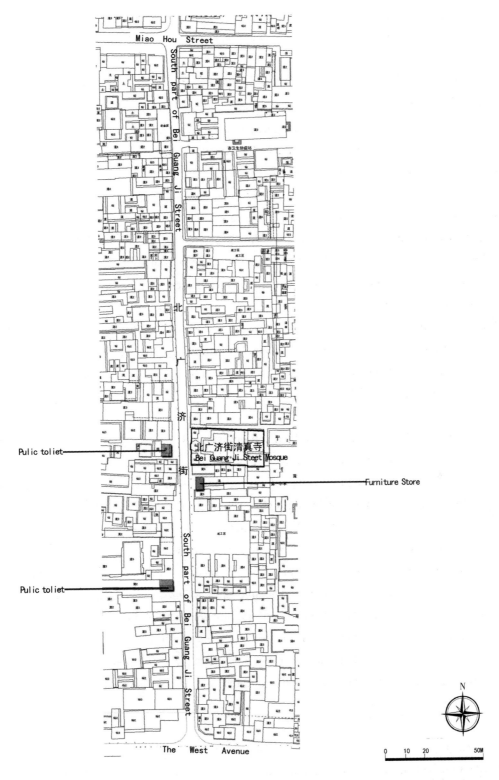

Figure 5-17 Public service along Guangming Alley and Beiguangji Street (southern part)

6) *Dapiyuan Street and Xiaopiyuan Street*

Both Dapiyuan Street and Xiaopiyuan Street are neighborhood level streets and their lengths are both approximately 210 meters. They connect the Bei Guangji Steet and Mai Xian Street. The two streets are both named as Pi Yuan, which literally means the courtyard where leather products are made. They were both constructed in Ming Dynasty and famous due to the prosperous transactions of leather products. To distinguish between two streets, the street at the north were named as Xiao (Chinese meaning is small) Piyuan Street, while the one at the south were named as Da (Chinese meaning is large) Piyuan Street.

Nowadays, in each street, there is one mosque named after the street. Therefore, there are many Muslims living around the mosques in these two streets. Being different with Dapiyuan Street and other streets in this study, the current main function of Xiaopiyuan Street is residential. Only one or two restaurants can be found at the ends of the street. Based on the observation, Xiaopiyuan Street locates near the edge of Muslim Quarter so that not many tourists would like to visit it. On the other hand, although it is not far from Xiaopiyuan Street, Dapiyuan Street is one of the most famous snack streets in Muslim Quarter. Except for a few residential buildings, public facilities and community services offices, most parts of the street are occupied by catering restaurants (Figure 5-18).

Xiaopiyuan Street can be considered as a non-commercial street (Figure 5-18, 5-19). Nonetheless, the inhabitants are facing the shortage of public services and community services. Currently, except for a few restaurants and shops, there are only three buildings that can be classified as public buildings, including a philharmonic orchestra for children, a clinic, and a gas station. However, in Dapiyuan Street, which is a popular snack street, there are many public facilities and community services. Figure 5-20 shows the layout of public buildings, from which we can see that there are quite a lot of groceries as well as other community services, including a pharmacy, a clinic, a vegetable shop, a hardware shop, and a garage and so on. Meanwhile, for education facilities, there is a kindergarten and a primary school in this street. Besides, there are some facilities with the purpose to serve the shop owners, such as a kitchenware store, a spice shop and a wholesale department. In addition, a car park is also in the street in order to serve both the visitors and residents, which is renovated from an obsolete office building's courtyard.

7) *Summary*

Based on the discussions above regarding the function of these streets in Muslim Quarter, some points could be concluded. First of all, in the past decades, along with the development of urban tourism and rapid commercialization, the function of the streets in Muslim Quarter changes

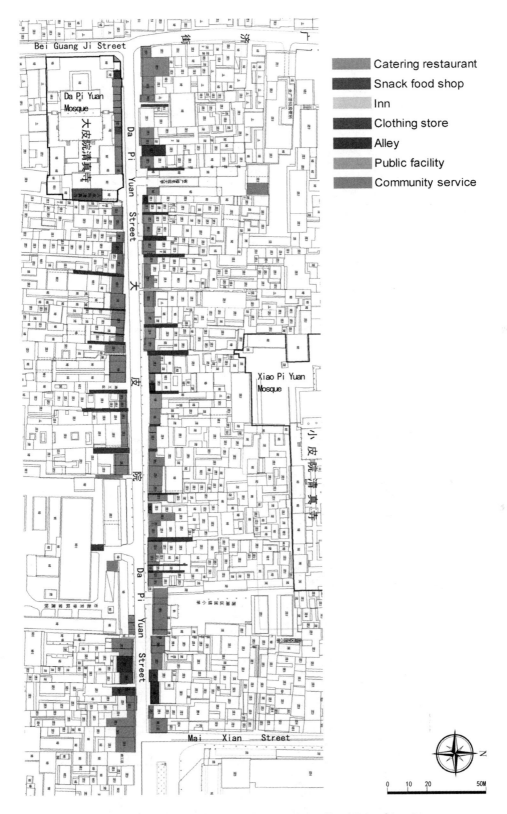

Figure 5-18　Functional layout along Dapiyuan Street

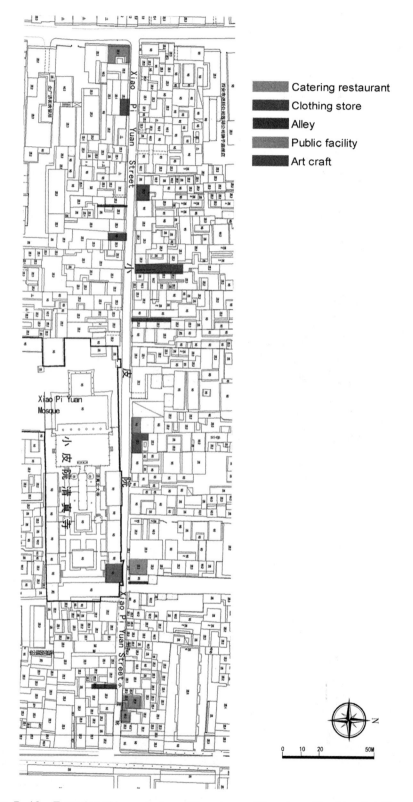

Figure 5-19　Functional layout along Xiaopiyuan Street——Residential Street

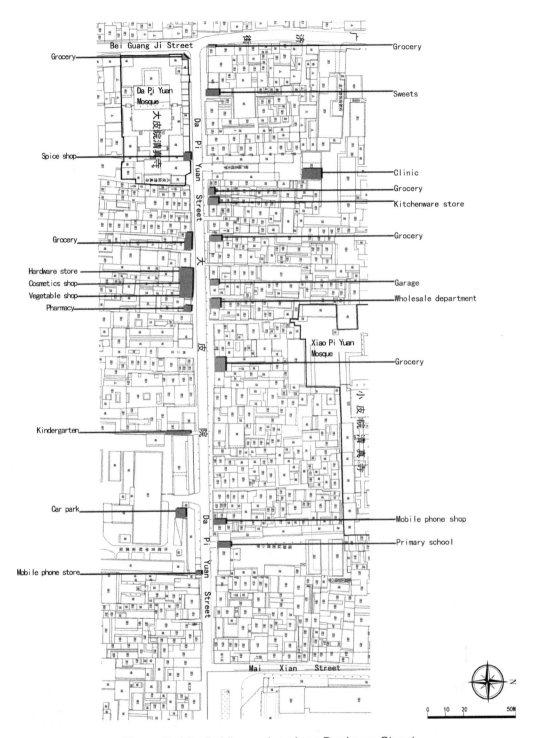

Figure 5-20　Public service along Dapiyuan Street

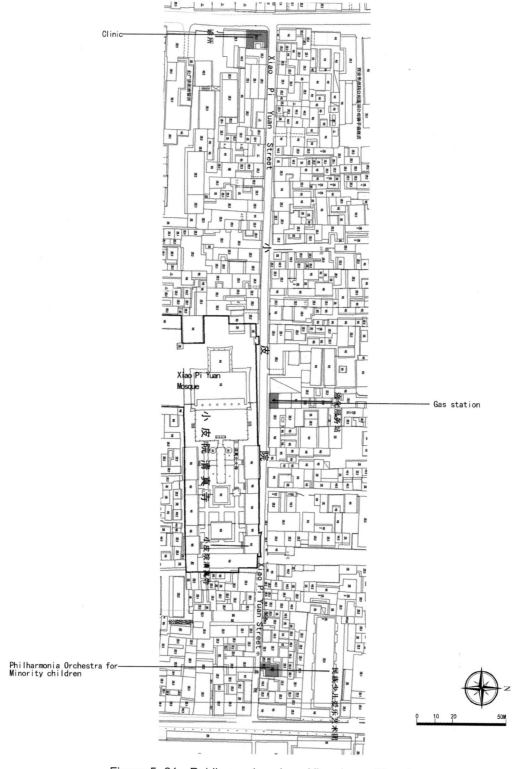

Figure 5-21　Public service along Xiaopiyuan Street

dramatically. Taking Xiyangshi as an example, compared to Lv's fieldwork in 1997 on the functional structure of the buildings along the street (Lv 1997), it shows that the number of residents has decreased sharply, from 32.8% to 1.6%. On the other hand, the retail section is increasing rapidly, from 28.60% to 52.80%. Secondly, most of the streets in Muslim quarter are mix used space. Xiaopiyuan is an exception among all cases, in which most of the buildings remain residential currently. In spite of this, there exist some public and commercial functions. Other streets are all fully mixed by various functions. Thirdly, the streets in Muslim Quarter could be considered as traditional space. Quite a lot of mosques are still playing very important roles to shape the streets' functional compositions. In addition, it is noticeable that streets among the entire community support each other functionally; hence the mapping of these streets are necessary in order to reveal among this community how the continuous social interactions in everyday pattern could contribute a stable social relationship in the long run. Nevertheless it has to be mentioned that the prosperous commercial activities in Muslim quarter might cause in serious traffic problems to the current streets system and affect the stability of functional structure upon these streets.

5.4.2 Street observation of Xiyangshi Street

In order to reveal the street life and social activities of local residents and to analyze how people use the street space, Xiyangshi Street was chosen as a typical example to carry out street observation. In this observation, the activity of shop owners, tourists and residents, the means of transportation in four days were recorded, which are 20-9-2012 (Thursday), 22-9-2012 (Saturday), 24-9-2012 (Monday), and 9-10-2012 (Tuesday).

1) *Transportation*

In Muslim quarter, the traffic congestion is quite a problem due to narrow streets and the poorly connected road system. With hundreds of year's history, the network of this district was not designed for modern transportation.Most of the people would choose walking, pedicab, bicycle, motorbike taxi and private vehicles. As shown in charts 5-1, 5-2, 5-3 and 5-4, in the mornings before 9:00am and evenings after 9:00pm, there were more people on the street and people would generally choose to walk and use pedicab. It is assumed that most of the people on the street in the early morning are local residents and shop owners and most of them on the street after 9:00pm are tourists because Xiyangshi Street is one of the most famous night markets in Xi'an (Figure 5-22, 5-23). In Chart 5-3 and 5-4, peaks also can be found in afternoons on weekday and quite a lot of people used motorbikes and taxis.

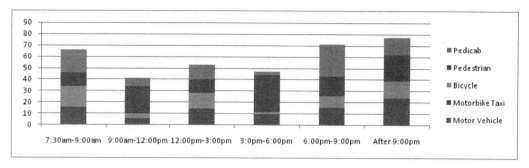

Chart 5-1　Transportation activities of Xiyangshi Street, 20-9-2012 (Thursday)

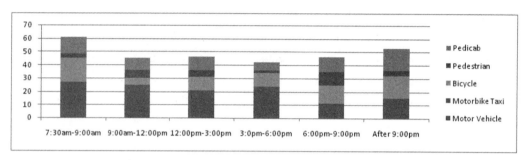

Chart 5-2　Transportation activities of Xiyangshi Street, 22-9-2012 (Saturday)

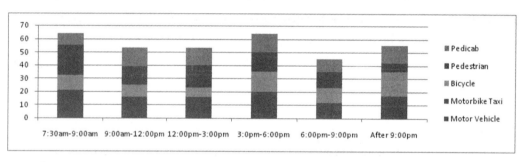

Chart 5-3　Transportation activities of Xiyangshi Street, 24-9-2012 (Monday)

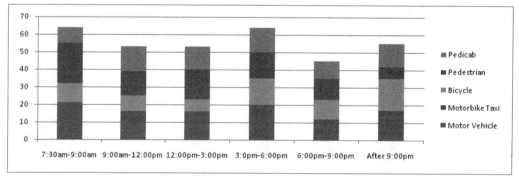

Chart 5-4　Transportation activities of Xiyangshi Street, 9-10-2012 (Tuesday)

Figure 5-22   Pedestrian use in the early morning

Figure 5-23   Traffic congestion

2) *Street activities*

Shop owners

Chart 5-5, 5-6, 5-7 and 5-8 summarize the activities of shop owners in four selected days. Overall, their activities were mostly observed during afternoon and evening. Meanwhile it is quite obvious that the most common activity is chatting with each other, which may prove that the shop owners are familiar with each other and their relationships are close. In addition, in the mornings, preparing to open their shops and opening their shops were frequently observed. Similarly, closing their shops were also the common activities after 9:00pm. In addition to these, stocking, baby setting, eating, and reading newspaper were also found on the street (Figure 5-24 to 5-28).

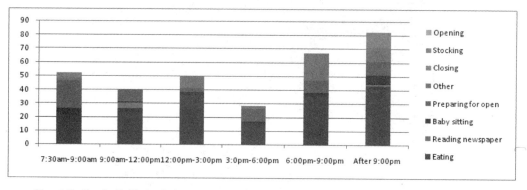

Chart 5-5   Activities of shop owners in Xiyangshi Street, 20-9-2012 (Thursday)

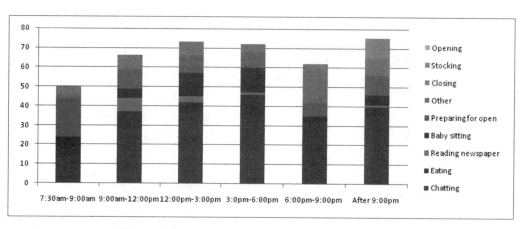

Chart 5-6   Activities of shop owners in Xiyangshi Street, 22-9-2012 (Saturday)

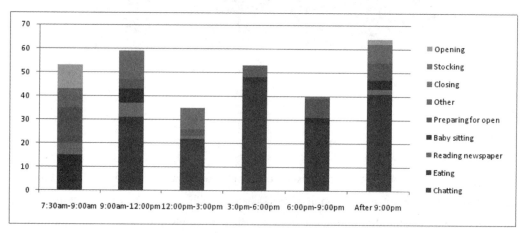

Chart 5-7   Activities of shop owners in Xiyangshi Street, 24-9-2012 (Monday)

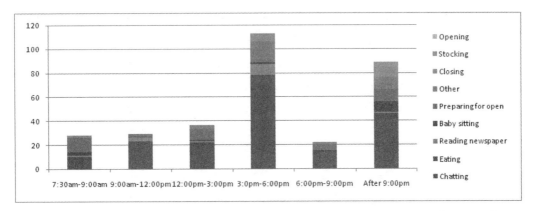

Chart 5-8 Activities of shop owners in Xiyangshi Street, 9-10-2012 (Tuesday)

Figure 5-24 Preparing food

Figure 5-25 Baby-sitting   Figure 5-26 Chatting

Figure 5-27 Giving in charity   Figure 5-28 Street vendors

Tourist

The activities of tourists were quite sample compared with residents and shop owners. Normally, three main activities could be observed on the street, including eating and drinking, shopping, and visiting. Besides, in the early morning before 9:00 am and late evening after 9:00 pm, there were fewer tourists on the street (Figure 5-29, 5-30).

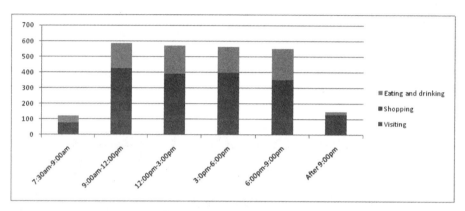

Chart 5-9　Activities of tourists in Xiyangshi Street, 20-9-2012 (Thursday)

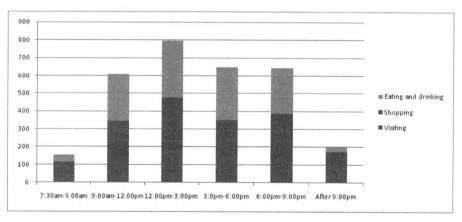

Chart 5-10　Activities of tourists in Xiyangshi Street, 22-9-2012 (Saturday)

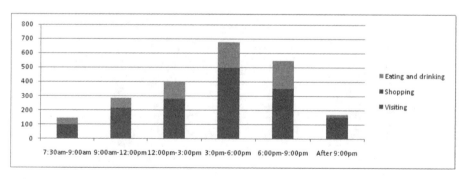

Chart 5-11　Activities of tourists in Xiyangshi Street, 24-9-2012 (Monday)

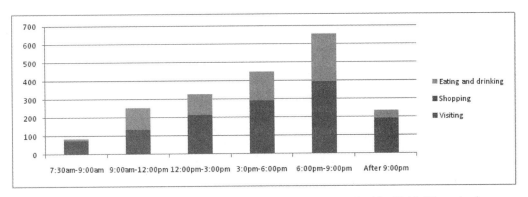

Chart 5-12 Activities of tourists in Xiyangshi Street, 9-10-2012 (Tuesday)

Figure 5-29 Shopping

Figure 5-30 Dining

Residents

As shown in Chart 5-13, 5-14, 5-15, and 5-16, residents' activities in Xiyangshi Street are different from that of shop owners' and tourists', including morning exercising, cycling, transporting items, going to public bath, walking children to and from school, children playing, strolling, eating, babysitting, grocery shopping, buying breakfast, walling birds, reading newspaper, chatting, watching, and sunbathing and so on. Especially, on the morning of 22nd September 2012 (Saturday), more than twenty people were preparing for a traditional Muslim Wedding on street. It is quite obvious that streets play a very important role in residents' social life. Among these activities, chatting with each other is the most common one could be found during the observation, which happened on various parts of the day. In addition, grocery shopping, babysitting, and watching were also quite frequently observed (Figure 5-31 to 5-37). Besides, some activities were more common on specific time, such as morning exercising and school commuting during 7:30 to 9:00am, sunbathing during 7:30-12:00am, and walking bird during 3:00 to 6:00pm.

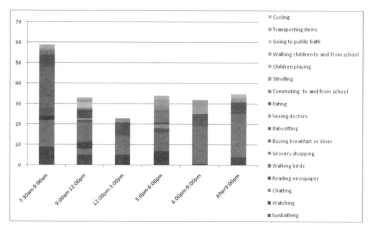

Chart 5-13　Activities of residents in Xiyangshi Street, 9-20-2012 (Thursday)

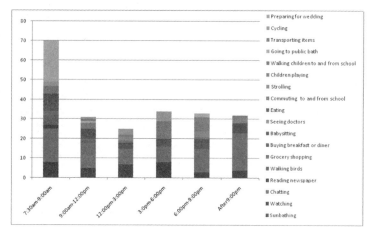

Chart 5-14　Activities of residents in Xiyangshi Street, 9-22-2012 (Saturday)

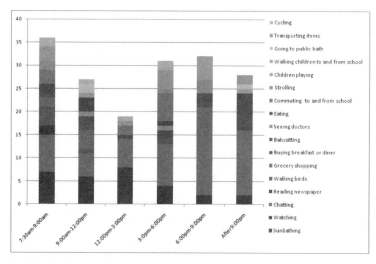

Chart 5-15　Activities of residents in Xiyangshi Street, 9-24-2012 (Monday)

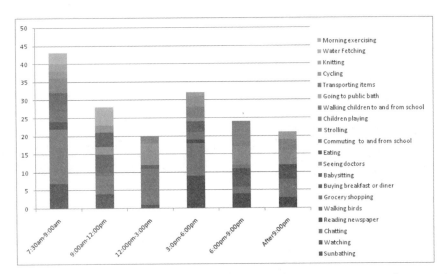

Chart 5-16　Activities of residents in Xiyangshi Street, 10-09-2012 (Tuesday)

Figure 5-31　Cricket fight gambling

Figure 5-32　Chatting

Figure 5-33　Preparing food for gathering

Figure 5-34　Street watching

Figure 5-35　Funeral

Figure 5-36　Wedding

Figure 5-37　Religious festival

3) Summary

In conclusion, this observation shows that there is quite rich street life in this neighborhood and street space is a very important component to build the sense of place. Social interactions like chatting and helping with each other are easily found in streets, which mean the social relationship between residents is very close. In addition, it also could be observed that the small scale commercial activities are very active during days and nights in weekdays and weekends.

5.4.3　Survey

To grasp the whole image of Hui Fang residents' attitudes towards the conservation and transformation of Hui Fang, 178 residents of two selected neighborhoods were interviewed, respectively 87 residents in Sajinqiao Street and 91 residents in the Great Mosque area.

1) *Demographic profile of respondents*

As can be seen in table 1, among the interviewees, 94.4% (168) of them are Muslims and only 10 residents (5.6%) assert themselves are non-religious. Meanwhile, 87.1% (155) of them are Ethnic Hui, 9.6% (17) are Han ethnic, and 3.4% (6) are Uygur. The statics indicates that in Hui Fang area, overwhelming majority of the residents are Ethnic Hui and believe in Islam. Another fact revealed by this data is that young people under 30 are relatively few in numbers; on the other hand, more than half of the interviewees are above 50 years old (Table 5-1).

Demographic profile of respondents — Table 5-1

		SAJINQIAO		THE GREAT MOSQUE		Total	
		N=87		N=91		N=178	
Gender	Female	33	37.9%	52	57.1%	85	47.8%
	Male	54	62.1%	39	42.9%	93	52.2%
Age	18-29	7	8.0%	14	15.4%	21	11.8%
	30-39	11	12.6%	3	3.3%	14	7.9%
	40-49	27	31.0%	22	24.2%	49	27.5%
	50-59	21	24.1%	24	26.4%	45	25.3%
	60-69	9	10.3%	21	23.1%	30	16.9%
	≥70	12	13.8%	3	3.3%	15	8.4%
Ethnic Group	Hui	72	82.8%	83	91.2%	155	87.1%
	Han	11	12.6%	6	6.6%	17	9.6%
	Uygur	4	4.6%	2	2.2%	6	3.4%
	Others	0	0.0%	0	0.0%	0	0.0%
Religion	Islam	79	90.8%	89	97.8%	168	94.4%
	Buddhism	0	0.0%	0	0.0%	0	0.0%
	Christianity	0	0.0%	0	0.0%	0	0.0%
	None	8	9.2%	2	2.2%	10	5.6%

2) *Socio-economic profile*

Now let's see Table 5-2. Regarding their educational level, more than half of interviewees in Sajinqiao Street are university graduates or possess post graduate qualifications, and in the Great Mosque area, 41 (45.1%) interviewees' education level is preliminary school and middle school, and only 4 people had graduated from university. Regarding their occupation, it can be seen that about one third (59 people) of them are retirees and there are also about one third (57 people) interviewees are self-employed.

Most of the respondents (81.5%) have lived in Hui Fang for more than 20 years, and over 90% of them have settled here for more than 10 years. Meanwhile, data shows that 86% of the respondents' flats are private property. About 9% live in public houses, and 5.1% rent the flats from other owners. Approximate half (48.3%) of the interviewees' families have four to five members, and about one third of them have more than 5 people in their families. Regarding their living space, the survey shows that 32% of the respondents' flats are between 50 and 90 square meters, while 64 (36%) interviewees' flats are bigger than 90 square meters but less than 150 square meters. This survey also includes the household income per year per capital which shows that there are 35.4% of the respondents who make money less than 4,500 CNY per year, 27.5% of them earn an annual income between 10,000 and 20,000 CNY, and only 3 interviewees' annual incomes are more than 50,000 CNY as show in Table 5-2.

Socio-economic profile   Table 5-2

		Sajinqiao N=87		The Great Mosque N=91		Total N=178	
Education level	None	5	5.7%	13	14.3%	18	10.1%
	Preliminary school and middle school	5	5.7%	41	45.1%	46	25.8%
	Technical secondary school and high school	30	34.5%	19	20.9%	49	27.5%
	Junior college Diploma	0	0.0%	14	15.4%	14	7.9%
	University and above	47	54.0%	4	4.4%	51	28.7%
Occupation	Student	4	4.6%	9	9.9%	13	7.3%
	Ordinary Worker	17	19.5%	5	5.5%	22	12.4%
	Technician	13	14.9%	6	6.6%	19	10.7%
	Manager	5	5.7%	3	3.3%	8	4.5%
	Self-employed	20	23.0%	37	40.7%	57	32.0%
	Retiree	28	32.2%	31	34.1%	59	33.1%
Length of residence	less than 3 years	2	2.3%	3	3.3%	5	2.8%
	3 – 10 years	3	3.4%	7	7.7%	10	5.6%
	10 - 20 years	9	10.3%	9	9.9%	18	10.1%
	over 20 years	73	83.9%	72	79.1%	145	81.5%
Ownership of flats	Private property	76	87.4%	77	84.6%	153	86.0%
	Directly managed old public housing	3	3.4%	6	6.6%	9	5.1%
	Rent from Danwei public housing	4	4.6%	3	3.3%	7	3.9%
	Rent from private owned flat	4	4.6%	5	5.5%	9	5.1%

		Sajinqiao		The Great Mosque		Total	
		N=87		N=91		N=178	
Flat area	less than 50m^2	3	3.4%	9	9.9%	12	6.7%
	50-90m^2	26	29.9%	31	34.1%	57	32.0%
	90-150m^2	35	40.2%	29	31.9%	64	36.0%
	150-200m^2	9	10.3%	7	7.7%	16	9.0%
	over 200m^2	14	16.1%	15	16.5%	29	16.3%
Family size	1-2 people	7	8.0%	9	9.9%	16	9.0%
	3 people	11	12.6%	6	6.6%	17	9.6%
	4-5 people	40	46.0%	46	50.5%	86	48.3%
	5 and above	29	33.3%	30	33.0%	59	33.1%
Household income per year per capital	Below CNY 4500	25	28.7%	38	41.8%	63	35.4%
	CNY 4500-6000	15	17.2%	9	9.9%	24	13.5%
	CNY 6000-10000	10	11.5%	8	8.8%	18	10.1%
	CNY 10000-20000	25	28.7%	24	26.4%	49	27.5%
	CNY 20000-50000	10	11.5%	9	9.9%	19	10.7%
	Above CNY 50000	0	0.0%	3	3.3%	3	1.7%

3) *Attitudes towards mosques, community committees and facilities*

Interviewees were requested to answer related questions in order to get a better understanding of local residents' opinion on the function of mosques and community committees. Table 5-3 summarizes the results and it shows that 45.5% (81 people) of the interviewees visits mosques every day, 25.8% people visit mosques three or four times per week, which indicates mosques might play a very important role in the building of social network. In the column of reasons to visit mosque, which is a multi-choice question, the survey reveals that most of the people (89.9%) go to mosques due to their religious belief, and attending funerals of relatives and friends and giving prayers, religious festival gatherings, and religion education are also strong motivations for them to visit mosques. Meanwhile, it also found that 81.5% of the interviewees agreed that mosque is a place accommodating their day-to-day practices of religion and tradition. Regarding the function of community committees, most of the interviewees described it as informing the residents of policies from governments and arrange activities (73.6%), and helping residents solve problems and difficulties (73.6%). While 37.6% of them agreed that it is poorly functional and not as good as mosques. The data also shows that more than 60% of the interviewees believed that changing facilities like schools and hospitals may cause huge impacts on local residents' everyday life.

Survey about Hui Fang——mosque and community committee   Table 5-3

		Sajinqiao		The Great Mosque		Total	
		N=87		N=91		N=178	
Q1: Which one should be the west edge of entire Hui Fang?	to West City Wall	41	47.1%	46	50.5%	87	48.9%
	to Zaoci Alley	52	59.8%	21	23.1%	73	41.0%
	to SajinQiao	4	4.6%	17	18.7%	21	11.8%
	to "the great Hui Fang" in 1980s	19	21.8%	29	31.9%	48	27.0%
Q2: How often do you visit mosques?	everyday	39	44.8%	42	46.2%	81	45.5%
	3-4 times per week	22	25.3%	24	26.4%	46	25.8%
	once a week	14	16.1%	12	13.2%	26	14.6%
	once a month	5	5.7%	4	4.4%	9	5.1%
	never	7	8.0%	9	9.9%	16	9.0%
Q3: Reasons to visit mosques.	Belief	79	90.8%	81	89.0%	160	89.9%
	Habit	36	41.4%	25	27.5%	61	34.3%
	Gathering with acquaintance gives a sense of belongings	22	25.3%	26	28.6%	48	27.0%
	Attend funerals of relatives and friends and give prayers	47	54.0%	61	67.0%	108	60.7%
	Muslim Festival gathering	56	64.4%	55	60.4%	111	62.4%
	local community congregation for community issues	34	39.1%	12	13.2%	46	25.8%
	Festival gala	2	2.3%	3	3.3%	5	2.8%
	Religion school	35	40.2%	58	63.7%	93	52.2%
	Community activity organized by community committee	5	5.7%	19	20.9%	24	13.5%
	Out-door exercise after gathering	19	21.8%	24	26.4%	43	24.2%
Q4: Do you agree the following statements about mosques (multi-choices)?	they are soly religious places	28	32.2%	21	23.1%	49	27.5%
	their religious functions are increasing	47	54.0%	33	36.3%	80	44.9%
	their social functions are increasing	49	56.3%	36	39.6%	85	47.8%
	the everyday life places	76	87.4%	69	75.8%	145	81.5%
	their influences to Muslim residents have declined	9	10.3%	11	12.1%	20	11.2%
Q5: How do you think community committee's function (multi-choices)?	inform the residents of policies from governments and arrange activities	63	72.4%	68	74.7%	131	73.6%
	help residents solve problems and difficulties	31	35.6%	74	81.3%	105	59.0%
	help families in poverty	19	21.8%	23	25.3%	42	23.6%

		Sajinqiao		The Great Mosque		Total	
		N=87		N=91		N=178	
Q5: How do you think community committee's function (multi-choices)?	improve community environment	36	41.4%	26	28.6%	62	34.8%
	organize community meeting on mutual community issues	8	9.2%	19	20.9%	27	15.2%
	poor function, not as good as Mosques	34	39.1%	33	36.3%	67	37.6%
Q6: How are the impacts of changes like schools and hospitals in Hui Fang on local residents' everyday life?	extremely huge	61	70.1%	49	53.8%	110	61.8%
	considerably large	16	18.4%	21	23.1%	37	20.8%
	some	17	19.5%	16	17.6%	33	18.5%
	few	3	3.4%	4	4.4%	7	3.9%
	none	0	0.0%	0	0.0%	0	0.0%

4) *Tradition*

This survey also shows that local residents values tradition very much. 65.7% of the interviewees believe it is very necessary to keep tradition for Muslim people and 86% of them believe that living in Hui Fang is extremely important for them to keep tradition and enhance their belief (Table 5-4).

Survey about Tradition   Table 5-4

		Sajinqiao		The Great Mosque		Total	
		N=87		N=91		N=178	
Q1: Do you think whether it is necessary to keep tradition for Muslim People?	not necessary at all	0	0.0%	2	2.2%	2	1.1%
	not very necessary, but can keep few	4	4.6%	5	5.5%	9	5.1%
	necessary, but allow positive changes	27	31.0%	23	25.3%	50	28.1%
	very necessary, tradition can be changed	56	64.4%	61	67.0%	117	65.7%
Q2: How does living in Hui Fang contribute to keeping tradition and enhancing belief?	extremely	79	90.8%	74	81.3%	153	86.0%
	considerably	8	9.2%	8	8.8%	16	9.0%
	some	0	0.0%	3	3.3%	3	1.7%
	little	0	0.0%	6	6.6%	6	3.4%
	none	0	0.0%	0	0.0%	0	0.0%

5) *Residents' perception towards Hui Fang's change and business activities*

Regarding the biggest change of Hui Fang in recent five to ten years which is a multi-choice question, the decline of historical remains and better business due to vertical growth of housing are represented in the most interviewees' opinions. Besides, 38.8% of them believed that their living standards have been largely improved. In addition, daily travel, sewage and garbage disposal, and cooking oil fumes are the top three disturbing matters for local residents due to Hui Fang's change and business activities as shown in Table 4-7. It also shows that more than a half of the interviewees would support the catering business and tourism in Hui Fang, and another 36% of them take the strongly supportive attitude to them (Table 5-5).

Residents' perception towards Hui Fang's change and business activities   Table 5-5

		Sajinqiao		The Great Mosque		Total	
		N=87		N=91		N=178	
The biggest change of Hui Fang in recent 5-10 years (multi-choice)	Business gets better with vertical growth of housing	43	49.4%	66	72.5%	109	61.2%
	The living standards of residents have been largely improved	28	32.2%	41	45.1%	69	38.8%
	Historical remains declined	41	47.1%	73	80.2%	114	64.0%
	Fewer Hui residents and more tourists	26	29.9%	32	35.2%	58	32.6%
	Traditional Hui Fang are disappearing	7	8.0%	15	16.5%	22	12.4%
The impacts to residents are (multi-choice)	Tourists activity affect privacy and everyday life	23	26.4%	54	59.3%	77	43.3%
	Affect the communication between neighbors	4	4.6%	3	3.3%	7	3.9%
	Affect the daily travel	71	81.6%	68	74.7%	139	78.1%
	Cooking oil fumes affects health	35	40.2%	54	59.3%	89	50.0%
	Sewage and garbage disposal	57	65.5%	62	68.1%	119	66.9%
	Noises	14	16.1%	31	34.1%	45	25.3%
	Affect children's secure play	16	18.4%	23	25.3%	39	21.9%
	Constrain the activities for the elderly residents	37	42.5%	39	42.9%	76	42.7%
Residents' attitudes towards catering business and tourism in Hui Fang	strongly support	27	31.0%	37	40.7%	64	36.0%
	support	49	56.3%	44	48.4%	93	52.2%
	oppose	6	6.9%	6	6.6%	12	6.7%
	strongly oppose	5	5.7%	4	4.4%	9	5.1%
	no objection	0	0.0%	0	0.0%	0	0.0%

6) *Culture and sense of place*

The survey shows that 92.1% of the respondents considered Hui Fang as a place embedded in religious spirits, 79.2% of them believed it is a closely bonded place, and 64% of them acknowledged it as a historical and cultural quarter. In terms of the understanding of Hui Fang's culture (multi-choice question), 93.8% of the interviewees chose "religion and traditional life style", 91% of them chose "traditional mosque-neighborhood dwelling culture", 88.2% of them chose "the traditional Muslim food culture", and 77.5% of them chose "a famous tourist spot", and so on. This indicates that in residents' mind, traditional lifestyle and mosque are the most significant component of local cultural and sense of place. Meanwhile, 89.3% of the respondents agreed it is necessary that the tradition and culture of Muslim should be conserved in Xi'an. Regarding the question of what is the contribution of Hui Fang to Xi'an City (multi choice), the most frequent answer is conserving the culture of minority ethnic groups (83.7%), and improving the tourism attraction and promoting the conservation of historic districts and cultures are also frequent answers which respectively are 75.8% and 74.7% (Table 5-6).

Culture and sense of place — Table 5-6

		Sajinqiao		The Great Mosque		Total	
		N=87		N=91		N=178	
Q1: How do you feel Hui Fang (multi-choices)?	a place embedded in religious spirits	84	96.6%	80	87.9%	164	92.1%
	a closely bonded place	73	83.9%	68	74.7%	141	79.2%
	a historical and cultural quarter	61	70.1%	53	58.2%	114	64.0%
	a place where the livelihood business depends on	42	48.3%	46	50.5%	88	49.4%
	an ordinary dilapidated area to be redeveloped	19	21.8%	27	29.7%	46	25.8%
Q2: How do you understand the culture of Hui Fang (multi-choice)?	the traditional Muslim food culture	71	81.6%	86	94.5%	157	88.2%
	Traditional Mosque- neighborhood dwelling culture	85	97.7%	77	84.6%	162	91.0%
	Traditional neighborhood street local business culture	59	67.8%	52	57.1%	111	62.4%
	Religion and traditional life style	84	96.6%	83	91.2%	167	93.8%
	religious festivals	29	33.3%	20	22.0%	49	27.5%
	a famous tourist spot	67	77.0%	71	78.0%	138	77.5%
Q3: Do you think the tradition and culture of Muslim should be conserved in Xi'an?	very necessary	79	90.8%	80	87.9%	159	89.3%
	necessary	6	6.9%	7	7.7%	13	7.3%
	no objection	2	2.3%	4	4.4%	6	3.4%
	not necessary	0	0.0%	0	0.0%	0	0.0%
	strongly not necessary	0	0.0%	0	0.0%	0	0.0%

		Sajinqiao N=87		The Great Mosque N=91		Total N=178	
Q4: What do you think Hui Fang's contribution to the city (multi-choice)?	improve tourism attraction	61	70.1%	74	81.3%	135	75.8%
	promote exchange and communication for the city	44	50.6%	49	53.8%	93	52.2%
	conserve the culture of minority ethnic group	72	82.8%	77	84.6%	149	83.7%
	promote the conservation of historic districts and culture	69	79.3%	64	70.3%	133	74.7%
	Function as "kitchen" of city, provide delicious food supply	26	29.9%	53	58.2%	79	44.4%

5.5 A general introduction of urban redevelopment in Hui Fang after 1980s

After 1949, as it is the case elsewhere in the country, most of the house properties in Hui Fang of Xi'an were socialized and redistributed to work units in a variety of ways. After reform and opening-up policy, a part of house properties were returned to their original owners, but there are still some houses managed by local authorities of housing administration and other work units (Cao 2005). Through long-term out-of-order transformation, it is quite normal that there are currently multiple families (from same clans for most of cases) living in the courtyard houses which originally were designed for one family (Dong 1996). For instance, No. 77 courtyard in Xiyangshi Street, as shown in Figure 5-38, which was bought by Ma Zijian (马子健) for his own family; in the second generation, this house was divided into four families by his children; and then in the third generation, it was further divided into ten households (Bi 2004).Consequently, lack of living space is a common problem in Hui Fang.

Staring in 1980s, the large-scale and rapid urbanization began and Hui Fang was also transformed in this process. According to Huang (Huang 2010), conflicts can be found between the redevelopment of Ethnic Hui's community and rapid urbanization. Generally, there are three major ways to conduct the redevelopment of traditional neighborhoods in Xi'an in accordance with the source of investments, namely local governments, private developers, and local residents themselves (Dong 1996). According to Dong (Ibid.), if the reconstruction is invested by local governments and private developers, the traditional neighborhoods would be greatly changed in terms of urban fabric and social structure because they are typical top-down approach with overall redevelopment strategies. Especially invested by private developers, the function of whole neighborhoods might be changed in order to get maximum investments returns. On the other hand, it could be considered as bottom-up approach if the reconstruction and conservation are invested, conducted (or supervised) by inhabitants themselves (Figure 5-39). Fortunately, the large-scale redevelopments in Hui Fang were not successfully carried out due to the collective action of conservation from local Hui people.

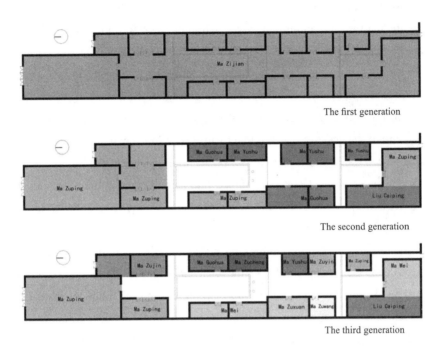

Figure 5-38 The subdivision of courtyard house within three generations
Source: author edited according to (Bi 2004).

Figure 5-39 Self construction by local residents in Huajue Alley

145

According to an unpublished research done by Dong and Wang[①], although the large scale reconstruction has not been carried out in Hui neighborhood in Xi'an City, small-scale reconstructions are quite commonly conducted by inhabitants in order to meet the need of Ethnic Hui families (Figure 5-38 and 5-39). According to this report, this bottom-up redevelopments were normally conducted inside the courtyard by inhabitants, therefore the urban fabric and social structure of the whole neighborhood remains undestroyed. By this approach, the inhabitants normally played the roles of designers, constructors and users so that the transformations could reflect their redevelopment wills for the Hui Fang neighborhoods. Dong and Wang believe that this approach could be considered as a way that HF neighborhoods evolve themselves to adapt to new circumstances and it is the main cause why the social structure and urban fabric that consists of streets, alleys and courtyards are successfully conserved in the process of rapid urbanization. However, Dong (1996) mentioned that some redevelopment without the professional assistance could not meet the construction standard due to the inhabitants' limited knowledge regarding architecture and civil engineering. Meanwhile, it is also a concern that some of the un-systemic expansions, especially those after 1990s, might undermine the streetscape of HF neighborhoods due to the weak control from the local authorities (Ibid.). For instance, some buildings were increased from 2 or 3 stories to 5-6 stories in order to acquire more space for commercial business and leasing. Figure 5-40 shows the transformation of street elevations of Xiyangshi Street from 2000 to 2012, which can demonstrate the increase of buildings heights and some of them had already broken through the height limit of 9 meters.Meanwhile, during the fieldwork in Hui Fang, it was found that some buildings were short of water supply and drainage system because the construction done by residents were without technical supports from professional architects and civil engineers. Therefore, around 2000s, local governments and organizations started getting involved into these small scale redevelopments in Hui Fang through supporting financial and technical supports. The details will be introduced in next section.

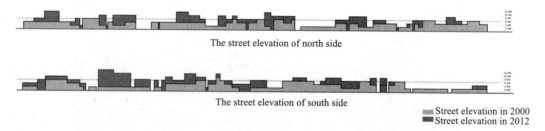

Figure 5-40 The transformation of street elevations of Xiyangshi Street from 2000 to 2012

---

① Research of housing reconstruction methods in Hui communities, Xi'an (西安回民区住宅改造方法研究) done by Dong and Wang, please refer to http://www.2muslim.com/forum.php?mod=viewthread&tid=2857.

Figure 5-41　The self-construction along Daxuexi Alley

Figure 5-42　The self-construction between North Guanji Street and Chenghuang Temple

5.6　The conservation process in Hui Fang

Since 1949, this district experienced self-construction, courtyard subdivision, and small-scale replacements of work-units, which lead to serious physical degradation.In 1980, Beiyuanmen historic street within the Muslim District was designated as a protected area (XCPB 1980).The conservation activities within this Muslim District can be examined in three stages: the initial stage, the transitional stage, and the target stage.

5.6.1 The initial stage (1980—2000)

The initial stage witnessed the ending of the Cultural Revolution, the recovery of social and economic activities, and the beginning of reform and opening-up. It is under such socioeconomic background that urban conservation is firstly mentioned in the Master Plan1980-2000 to enhance the historical features and townscape of Xi'an. In Table 5-7, the culture relics and buildings at all level (national, provincial and municipal) in Hui Fang are listed.

Culture relics and buildings in Hui Fang　　　　　Table 5-7

Number	Level	Culture relics and buildings	Dynasty
1	National	The great Mosque（西安清真寺）	Ming
2		Bell tower and Drum Tower（西安钟楼、鼓楼）	Ming
3		Chenghuang Temple（西安城隍庙）	Ming and Qing
4	Provincial	Xiaopiyuan Mosque（小皮院清真寺）	Ming
5		Daxuexi Alley Mosque（大学习巷清真寺）	Ming and Qing
6	Municipal	Dapiyuan Mosque（大皮院清真寺）	Ming
7		Courtyard 144 in Beiyuanmen Street（北院门144号民居）	Ming and Qing

In the Master Plan, Beiyuanmen Historic Street (Figure 5-43), a prominent historic street within the Muslim District, together with another inner city street, were designated as two historic neighborhood streets. The Master Plan changed from the previous single physical spatial planning approach (XCPB 1980). Due to the lock-in effect of previous policies under a planned economy, during the initial stage, conservation practices continued to adopt short-term single physical revitalization strategy with a top-down process. Aside from the awareness of tourism-led conservation, the initial stage also saw collaboration with overseas institutions and foreign funding as emerging trends. The initial stage of conservation is observed in the Xi'an Muslim Historic District Protection Project.

In 1997, the "Sino-Norwegian Cooperative Plan for the Protection of Xi'an Muslim Historical District" (XMHDPP) was formed by the central government of China. Subsequently, an official sub-district office, the Xi'an Muslim Historic District Protection Project Office, was established to execute the implementation of the XMHDPP. During the implementation of this project, concentration was placed primarily on physical upgrading, the improvement of infrastructure and the environment, and the restoration of three traditional courtyard houses. In this project, cer-

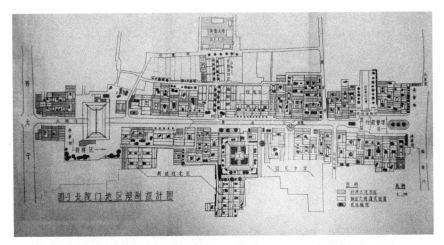

Figure 5-43　Planning of Beiyuan men Street, 1983
Source: Mr.Ma Yunhai.

tainconservation methods were adopted and the results are satisfactory. Three cases are worthy to be discussed, which are courtyard house No.77 in Xiyangshi Street, courtyard house No. 125 in Huajue Alley, and courtyard house No. 144 in Beiyuanmen Street.

Courtyard 77 in Xiyangshi Street has been introduced in the previous section, which is the estate of Ma family. Due to the increase of family members, this courtyard was extremely crowded. Therefore, besides the renovating the building structure and renewing the infrastructure, the experts of this project reorganized the shortage of living space during the conservation. Interlayer was added between first floor and second floor without changing the facades, which was designed as bedrooms (Bi 2004). The method solved the problem of scarcity of living space, but still some residents complained that the infrastructures were not sufficient (Ibid.).

Different with courtyard house No.77 in Xiyangshi Street, there is only one family lived in courtyard house No. 125 in Huajue Alley all the time. The main problem is that the building and courtyard were seriously damaged during the Anti-Japanese War by the Japanese aerial bombardment. Therefore, the conservation project repaired the structure and renovated the courtyard house according to its pre-war stage (Ibid.). The function of the house remains residential and there is no household change (Figure 5-44).

The third case is courtyard 144 in Beiyuanmen Street that was an office courtyard that belonged to Xi'an Chinese Painting Academy before conservation. In this project, this courtyard was restored according to its appearance in Qing Dynasty with a functional reuse, and the renovated courtyard house is now functioning as a tourism spot to demonstrate the history of this neighborhood (Ibid).

149

Figure 5-44　Renovated courtyard house No.125 in Huajue Alley

5.6.2　The transitional stage (2000—2008)

In 2002, the Xi'an Historic City Conservation Regulation (XHCCR) under Master Plan 1995-2010 was issued for implementation as law, in which the boundaries of the Beiyuanmen Historic District (BHD) were designated (Figure 5-45).

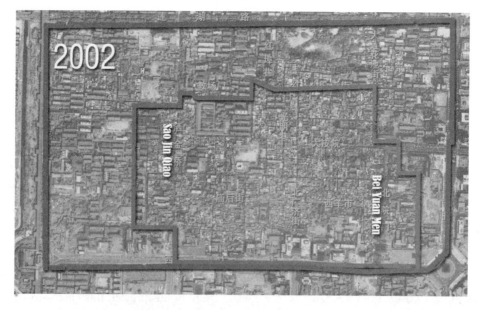

Figure 5-45　Scope of BHD in Xi'an Master Plan (1980—2000)
Source: Xi'an Master Plan 1980—2000.

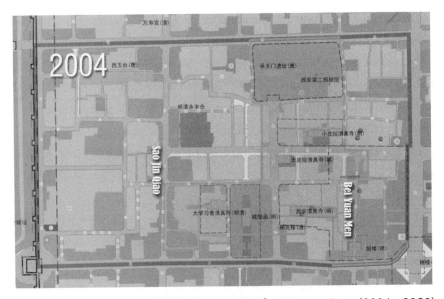

Figure 5-46　Scope of BHD in Proposed Xi'an Master Plan (2004—2020)
Source: Xi'an Master Plan (2004—2020).

The transitional stage saw deep social and economic transitions. With the process of transitioning from a planned economy to a market economy, institutional change favored the market-led economy and overlooked the function of the planned economy. Along with the decentralization of power from central to municipal government, real estate has been given priority in urban development. Although many historic quarters are located in central urban areas, redevelopment with mass demolition and relocation was widely adopted. Accordingly, this transitional stage witnessed the weakening of conservation regulation as local governments adjusted the scope of the pre-existing conservation plan to favor real estate. The redevelopment project on Sajinqiao Street within the Muslim historic district offers a reflection of this.

In 2004, Lianhu District initiated the Sajinqiao Redevelopment Plan (SRP). At the same time, the Xi'an City Planning Bureau licensed a land use permit (No.2004-289) of 98 acres for a road-widening project on Damaishi Street and Sajinqiao Street, which only require 20 acres. 70 acres of spare land where 1,900 houses were located of which 90% were Muslim residents, were demolished. The compensation price offered by Lianhu District was rather low compared to the market price. Both private developers and the Development Center of Lianhu District were involved in this project (Figure 5-46).

According to the proposed plan, many Muslim residents were to be relocated to suburban areas, separated from their traditional environment, religious space, and economic practices. Most of the residents who were refusing to move filed a petition to the Provincial Government. As a result, the first round of demolition was terminated. In April 2005 in the newly proposed Xi'an Master Plan 2004—2020 (XCPB 2004), the scope of the Beiyuanmen Historic District was adjusted by the Xi'an City Planning Bureau. Including both Sajinqiao Street and Damaishi Street, eight historic streets, which were previously designated in the BHD in XHCCR, were excluded. In the wake of this adjustment and fearing the re-emergence of mass demolition acts, residents of Sajinqiao Street, along with religious leaders, gathered at the West Mosque where resident representatives were elected and eleven Mosques were united inorder to conserve their community (Figure 5-47).

The central government intervened and the second round of demolition was stopped in July 2006. The feedback from the Xi'an Planning Bureau in June 2006 asserted that before the Master Plan 2004—2020 was finally approved by the central government, all redevelopment regarding the Muslim District should strictly refer to XHCCR 2002. In this transitional stage, local residents and Mosques actively participated in the conservation of a place, strengthening the weakening conservation regulation and enhancing their social, economic, and culture needs. This successfully challenged the profit-led redevelopment pattern and largely contributed to shaping new conservation planning for the following stage.

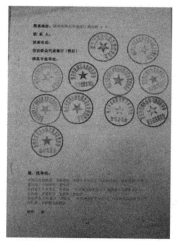

Figure 5-47 The Eleven mosques' joint seals to establish the Mosque Union and one page of local residents' joint signature in the petition to the central government against demolition

5.6.3 The target stage (2008—2020)

In June 2008, the new Xi'an Mater Plan 2008—2020 was approved by central government. The municipal government assured that the social, economic, and physical needs of local residents in Hui Fang would be given priority in future redevelopment plans. In the future, through an inclusive conservation planning approach, the Muslim District will adopt broader methods to manage changes, and, at the same time, ensure that the existing community will be socially, economically, and physically resilient in the long run.

5.7 A survey on local residents' perception towards Hui Fang's conservation

5.7.1 Residents' attitudes towards urban redevelopment

The residents' attitudes towards urban redevelopment are also included in this research, which is shown in table 4-6. It is quite clear that most preferable scale of redevelopment is small scale according to building quality. Towards the redevelopment projects in nearby areas, attitude of very negative accounts for 61.2% of all response, and the one of negative accounts for 17.4%. Only 13.5% of the interviewees believed they have positive impacts on urban redevelopment. Regarding the resettlement approach, 45.5% of the interviewees chose to stay in Hui Fang, 18.5% of them might decide based on compensation, and only 7.3% of them would take the initiative to move out of Hui Fang. Most of them concerned that the redevelopment could destroy the historic value of Hui Fang and weaken the close relationship within Muslim neighborhoods. Meanwhile, they also worried about the difficulty of visiting mosques after redevelopments.

5.7.2 Changes of life in recent 10 years

Table 5-8 shows the residents life change in recent 5 years. In it, it can be seen that more than a half of the interviewees stated that their incomes are increased and 18% of them complained their income are worsened than before. Besides, 53.9% of the respondents said that their purchasing power is improved, while 19.1% of them believed it is worse than before. Regarding their attitudes for income, the percentage for satisfactory is 30.3%; the one for non-objection is 29.2%; and the one for unsatisfactory is 25.8% as shown in Table 5-8.

This unsatisfactory situation of Sajinqiao Street could be explained by its continuous physical deterioration and functional decline after the top-down demolition project in 2007 (Figure 5-48). Though the municipal government claimed to provide an alternative renewal plan which will not relocate local people and will encourage public participation, till 2014 there has no such plan available. And no measures have been implemented to improve the physical environment and to revitalize the economic situation of this area. Due to this government "allowed" decay, many shops have to close up, and remaining residents have suffered from in terms of physical decay and economic loss. A comparison of the field survey in 2013 by this research and in 1997 by Lv (1997) could also provide insights regarding this issue.

Life change in recent 5 years   Table 5-8

		Sajinqiao		The Great Mosque		Total	
		N=87		N=91		N=178	
Income	Largely improved	2	2.3%	14	15.4%	16	9.0%
	Improved	37	42.5%	65	71.4%	102	57.3%
	Remain the same	21	24.1%	7	7.7%	28	15.7%
	Worsened	27	31.0%	5	5.5%	32	18.0%
Purchasing power	Largely improved	2	2.3%	4	4.4%	6	3.4%
	Improved	39	44.8%	57	62.6%	96	53.9%
	Remain the same	25	28.7%	17	18.7%	42	23.6%
	Worsened	21	24.1%	13	14.3%	34	19.1%
Income	Very satisfactory	3	3.4%	3	3.3%	6	3.4%
	Satisfactory	5	5.7%	49	53.8%	54	30.3%
	No objection	34	39.1%	18	19.8%	52	29.2%
	Unsatisfactory	30	34.5%	16	17.6%	46	25.8%
	very unsatisfactory	15	17.2%	5	5.5%	20	11.2%

Figure 5-48   Physical deterioration and functional decline of Sajinqiao Street

Aside from the economic aspects, interview with local residents and Mosque leaders reveal the growing concerns about the obsolescence of Huis' educational and medical facilities within Hui Fang, particularly the kindergardens and preliminary schools. During the last ten years, with the integration and reallocation of public resources such as medical and education in the municipal and district level, several Hui's neighborhood educational institutions which serve mostly the Hui's have been abandoned (Figure 5-49 to 5-54) and few remaining schools have satisfactory rankings, which left no much choices for the Huis and they have to receive education outside the neighborhoods together with the Hans. The differences in lifestyles, especially the dietary difference has caused inconvenience.

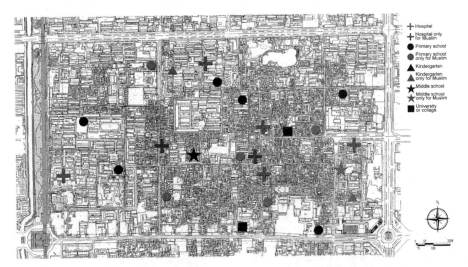

Figure 5-49 Medical and educational resources in HF in 1997
Source: Author edits according to Lv's research (1997).

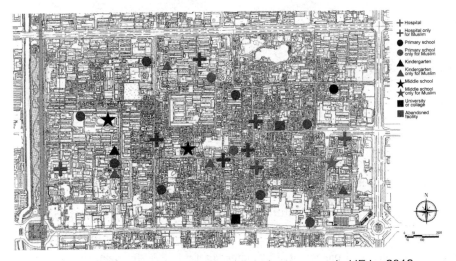

Figure 5-50 Medical and educational resources in HF by 2013

Figure 5-51 Abandoned Sajinqiao Preliminary School

Figure 5-52 Abandoned Huajuexiang Preliminary School

Figure 5-53 Abandoned Tuanjie Preliminary School

Figure 5-54 Abandoned Huis Hospital

5.7.3 Residents' attitude towards the Beiyuanmen Historic District Sino Cooperation Preservation Project (1997—2002)

In order to reveal local residents' attitude towards Beiyuanmen Historic District Sino Cooperation Preservation Project (1997—2002), a survey was conducted and 91 local residents in the Great Mosque area participated. As show in Table 5-9, fourteen people (15.4% of all respondents) took part in this project, among them only two residents' houses were selected as traditional courtyard and conserved. While 34 interviewees were just notified by the community committee and 39 interviewees had nothing to do with this project. Among all project participants, consultants from universities were the most active one in terms of contacting local residents. Street office, local police station and community committee also help a little bit to communicate with residents. But other participants, including the Great Mosque, Beiyuanmen District Preservation Project Office, the residents' representatives, and construction sector were not paying enough attention in contacting residents. About a quarter of the respondents believed that consultants from universities offered lots of help. When answering the multi-choice question of "What is the most important influence to you in this project?", 80.2% of the interviewees believed that this project help them to get a profound understanding of historic architecture and street, 52.7% of them supposed this project made Hui Fang more famous and would attract more tourists, and only 12.1% of them thought that their houses were renovated during this project. Overall, 79.1% of the interviewees believed this project enhanced the awareness of urban conservation.

Interview of residents' attitudes towards the Beiyuanmen Historic District Sino Cooperation Preservation Project (1997—2002)　　　　Table 5-9

			The Great Mosque	
			N=91	
Q1: Did you participate in Beiyuanmen Historic District Sino Cooperation Project 1997-2002)		Yes	14	15.4%
		No	77	84.6%
Q2: Which part did you participate?	Only was notified by the community committee about the project		34	37.4%
	Actively participated in the Preservation Project		0	0.0%
	Your own house was selected as traditional courtyards to be conserved		2	2.2%
	Your drainage pipeline was restored in this project		13	14.3%
	The façade of your house was decorated in this project		3	3.3%
	None		39	42.9%
Q3: How often did these following organizations contact you during the project?				
the Great Mosque	very close		2	2.2%
	close		7	7.7%
	seldom		62	68.1%
	never		20	22.0%
Community Committee	very close		13	14.3%
	close		32	35.2%
	seldom		29	31.9%
	never		17	18.7%

		The Great Mosque	
		N=91	
Street Office	very close	7	7.7%
	close	29	31.9%
	seldom	42	46.2%
	never	13	14.3%
Local Police Station	very close	11	12.1%
	close	5	5.5%
	seldom	58	63.7%
	never	17	18.7%
Beiyuanmen District Preservation Project Office	very close	3	3.3%
	close	21	23.1%
	seldom	25	27.5%
	never	42	46.2%
Consultants from universities	very close	21	23.1%
	close	17	18.7%
	seldom	34	37.4%
	never	19	20.9%
The residents representatives	very close	0	0.0%
	close	5	5.5%
	seldom	59	64.8%
	never	27	29.7%
Construction Sector	very close	3	3.3%
	close	10	11.0%
	seldom	66	72.5%
	never	12	13.2%
Q4: How much did these organizations help you			
the Great Mosque	lots of help	0	0.0%
	much help	8	8.8%
	little help	38	41.8%
	no help	35	38.5%
Community Committee	lots of help	8	8.8%
	much help	37	40.7%
	little help	34	37.4%
	no help	12	13.2%

			The Great Mosque	
			N=91	
Street Office		lots of help	14	15.4%
		much help	26	28.6%
		little help	43	47.3%
		no help	8	8.8%
Local Police Station		lots of help	13	14.3%
		much help	22	24.2%
		little help	41	45.1%
		no help	15	16.5%
Beiyuanmen District Preservation Project Office		lots of help	6	6.6%
		much help	17	18.7%
		little help	22	24.2%
		no help	46	50.5%
Consultants from universities		lots of help	23	25.3%
		much help	25	27.5%
		little help	27	29.7%
		no help	16	17.6%
the residents representatives		lots of help	0	0.0%
		much help	3	3.3%
		little help	57	62.6%
		no help	31	34.1%
Construction Sector		lots of help	5	5.5%
		much help	7	7.7%
		little help	63	69.2%
		no help	16	17.6%
Q5: What is the most important influence to you in this project (multi-choice)?		Have a profound understanding of historic architecture and street	73	80.2%
		Our houses were renovated	11	12.1%
		Infrastructure were upgraded	36	39.6%
		Community Environment were improved	45	49.5%
		HuiFang got more famous and attracted more tourists	48	52.7%
Q6: Do you believe that this project has enhanced your awareness of urban conservation?		Yes	72	79.1%
		No	19	20.9%

5.7.4 Collective actions towards the Sajinqiao Redevelopment Project (2005—2007)

A similar survey (Table 5-10) was conducted with the intention to understand residents' attitudes towards the Sajinqiao Redevelopment Project (2005—2007), which received 87 responses from the local residents in Sajinqiao Street. Difference between the Beiyuanmen Historic District and Sino Cooperation Preservation Project, is that there are 76 people who took part in Sino project, which makes up 87.4% of all respondents. According to this survey, residents actively took actions to resist the implementation of this project, such as writing a petition to local government, attending residents' meeting to discuss the conservation of entire Hui Fang and so on. In this project, mosques played a significant role and 59.8% of the respondents believed that they were closely contacted by mosques. Meanwhile, the residents' representative group and the volunteer group were also very active in communicating residents. Consequently, interviewees stated that mosques, the residents' representative group, and the volunteer group were very helpful. Almost all of the respondents believed that they got a profound understanding of the conservation of entire Hui Fang. 79.3% of them stated that they gained social justice through cooperation between neighbors. While 77% of them would contribute to Hui Fang's future development and encourage the conservation of entire Hui Fang.

Interview of collective actions towards the Sajinqiao Redevelopment Project (2005—2007)　　　　Table 5-10

			Sajinqiao	
			N=87	
Q1: Did you participate in the collective actions in response to Sajinqiao redevelopment project (2005-2007)		Yes	76	87.4%
		No	11	12.6%
Q2: Which part did you participate (multi-choice)?	Only was notified by the community committee about the redevelopment project		7	8.0%
	Actively participated in the conservation of Sajinqiao Neighborhood		53	60.9%
	Attended Residents' meeting and discussed the conservation of entire Hui Fang		68	78.2%
	Taken part in a petition to local government		71	81.6%
	Joined in the residents volunteer organization		13	14.9%
	Elected as Residents Representatives and dedicated to the conservation of this neighborhood		2	2.3%
	Elected as 11 Mosques Representatives and dedicated to the conservation of entire Hui Fang		0	0.0%
Q3: How often did these following organizations contact you during the process of collective resistance?				
Mosques		very close	52	59.8%
		close	31	35.6%
		seldom	4	4.6%

			Sajinqiao	
			N=87	
Mosques		never	0	0.0%
Community committee		very close	0	0.0%
		close	0	0.0%
		seldom	51	58.6%
		never	36	41.4%
Local police station		very close	0	0.0%
		close	2	2.3%
		seldom	28	32.2%
		never	57	65.5%
The residents representative group		very close	31	35.6%
		close	51	58.6%
		seldom	5	5.7%
		never	0	0.0%
The volunteer group		very close	37	42.5%
		close	30	34.5%
		seldom	17	19.5%
		never	3	3.4%
	Q4: How much did these organizations help you?			
Mosques		lots of help	64	73.6%
		much help	21	24.1%
		little help	2	2.3%
		no help	0	0.0%
Community Committee		lots of help	0	0.0%
		much help	6	6.9%
		little help	53	60.9%
		no help	28	32.2%
Local Police Station		lots of help	0	0.0%
		much help	4	4.6%
		little help	34	39.1%
		no help	49	56.3%
The residents' representative group		lots of help	25	28.7%
		much help	37	42.5%
		little help	14	16.1%
		no help	11	12.6%
The volunteer group		lots of help	31	35.6%
		much help	27	31.0%

			Sajinqiao	
			N=87	
The volunteer group	little help		19	21.8%
	no help		10	11.5%
Q5: What is the most important influence to you in this collective action to redevelopment (multi-choice)?				
Have a profound understanding of the conservation of entire Hui Fang			84	96.6%
Small scale business as the livelihood are secured			39	44.8%
Our houses are saved			46	52.9%
Started to trust the neighbors and would like to cooperate together in future			43	49.4%
One's own interest are secured and stayed in Hui Fang			54	62.1%
Will contribute to Hui Fang's future development and encourage the conservation of entire Hui Fang			67	77.0%
Gain social justice through cooperation between neighbors			69	79.3%

Chapter 6 Case Study: Tianzi Fang in Shanghai

6.1 A brief introduction of current historic district conservation planning in Shanghai

Shanghai was listed in the second group of National Famous Historical and Cultural Cities in 1986. Current conservation planning include five sectors which are: 1) historical and cultural buildings and outstanding modern architectures; 2) central historical and cultural areas; 3) historical towns; 4) landscape protection areas; 5) natural reserve areas.[①] Among these sectors, the conservation of historical and cultural districts is closely involved in the urban development process and plays an increasingly important role in promoting Shanghai's urban image and strengthening its competitiveness in a global stage (Zhang 2006). In 2002, Conservation Regulation Planning for Shanghai Historical and Cultural Areas was issued by the Shanghai Municipal Government, which designated twelve historic and cultural areas in order to conserve the historic and cultural features of central Shanghai.

6.1.1 Detailed regulatory planning for twelve areas with historical and cultural features

As shown in Figure 6-1, twelve areas with historical and cultural features were defined that covers total of 27 square kilometers in this planning document. The core area of this protection planning are the places with the well-conserved urban fabric, historical buildings and the peripheral construction-control areas, which are 339hm^2 and takes up 44% of the whole areas. Public buildings and residential buildings, including Li-Nong, villas and modern apartments were included in the conservation list. Except for some technical repairing for building structures and upgrading of indoor space, the basic conservation methods in larger scale include increasing green space and public space and decreasing plot ratio and total number of buildings. Meanwhile four major principles were mentioned; (1) Integrity principle which means to protect urban cultural legacies of Shanghai, especially the residential buildings and famous historical events related sites; (2) Authentic principle which means to protect the original spatial layout, urban fabric, street scale, and cultural diversity; (3) Sustainable principle which means to complete the functions and landscapes, improve living conditions, and use various protective

---

① Office of Shanghai History,
http://www.shtong.gov.cn/node2/node2245/node64620/node64632/node64720/index.html.

methods to make the historical sites meet the demands of modern life and improve their qualities; (4) Classification protection principle which means various methods will be carried out according to different conditions and cultural value of historical buildings in order to make sure the diversity of historical sites and the operability of this plan.Noticeably Tianzi Fang is not listed in the twelve areas.

Figure 6-1 Twelve historical and cultural areas

Source: Tongji University, Conservation Planning of Historical and Cultural City of Shanghai, 2002.

6.1.2 Urban conservation approaches and practices on historic quarters

With regards to the conservation approaches of historic quarters in Shanghai, along with the implementation of the conservation regulation planning, various approaches have been employed into the conservation practices, for instance, the exchange of property right in the conservation of Sinan Road Mansion; the historical environment restoration of Wukang Road; the adaptive reuse of industrial heritage and creative industry led revitalization in Shiliupu Old Pier. Nevertheless the followings three representative cases have provided influential conservation approaches for the conservation practices that followed.

The comparison of three cases　　　　　　　　　　　　　　　Table 6-1

	Xintiandi	Bugaoli	Moganshan Road M50
Location	Located in the eastern part of Huaihai Road, Luwan District	Located in the junction of south Shanxi Road and west Jianguo Road, Luwan District	Located at the southern bank of Suzhou River, Putuo District

	Xintiandi	Bugaoli	Moganshan Road M50
The original functions.	Mainly as residences	Residences	Industry
Type of architecture.	Old Shanghai Li-Nong	Old Shikumen Li-Nong	Li-Nong factories
Introduction	It was built in early twenties as residential Li-Nong neighborhood in French concession area	In 1930, it was developed by a French company for Chinese people, which is a typical residential Li-Nong neighborhood, composed by 78 timber-structure houses	In 1933, this family industry was built by a famous Anhui merchant Mr Zhou, and invested by foreign shareholders and domestic shareholders
Scale	The area of the planning site is 30,000m^2, and has 57,000m^2 of structure area. After renew, the total area of its north part is 24,800m^2, and its south part is 25,000m^2	The area of the site is about 7000m^2, in which there are 11 buildings and 79 units; the total gross floor area is about 10,069m^2	The total area of the site is 24,000m^2 and it has 41,000m^2 of historical factory architectures
Function transformation	Form residential function to commercial function (Guan 2008)	Keeping its original residential function	From industrial function to multi-function such as business, leisure culture, display and business services, etc (Huang 2008)
The renewal method in building- scale.	Based on the significance and status, three methods were adopted accordingly: 1. Preserving and reinforcing structures; 2. Remaining the outside walls and roofs and reconstruct inner structures; 3. Demolishing most parts of the buildings and only keep the gate houses and keeping the block at its original space scale	1. Updating kitchen facilities. 2. Installing advanced toilets. 3. Repairing outside walls. 4. Installing fire sprinkler	Adding the featured factory buildings into the Historical Building List and focusing on their preservation
Renewal methods in neighborhood- scale	Installing modern facilities, including drainage system and air conditioners	Keeping its original building appearances, block styles and residential functions; Developing tourism function appropriately. 1. Reconstructing Roads. 2. Repairing municipal pipe systems	1. To guide the industry, make whole area form as a modern culture art community, which mainly consists of artist studios, art exhibitions and conference center (Huang 2008). 2. Combining individual industrial buildings, production lines and factory, from a factory-featured exhibition and industry tourism cultural preservation. 3. Adding more living facilities, such as bars, shops, to make convenience to artists

	Xintiandi	Bugaoli	Moganshan Road M50
Business patterns.	After redevelopment, special shops were run by time-honored brand merchants, stars, and creative sellers in order to attract customers and tourists	Keeping original function as a residential neighborhood, which can be considered as a way to benefit local residents	65% were leased to galleries and studios, 35% are design companies, media companies, institutions of higher education, and so on
Transition of property rights.	Relocating out all of the original residents to realize function replacement and transformation	Most residents bought new houses and rent the old ones out. Rental income was an important way to make a living for residents. At present, among 900 residents here, 372 households are tenants, which could easily lead to high population density, improper use of the buildings, and the conflictions between tenants and local residents	Keeping the real estate property rights to make sure owners' profits. As a creative industry park, the nature of the land was still industry, and the property belongs to original enterprises. Investors and owners signed a lease agreement ranging from 5 to 10 years
Participants and operation pattern	Luwan District government cooperated with HongKong Ruian Group	It was a complete government action done by Luwan district government and Luwan District Housing and Land Administration	It was a folk spontaneous activity and later intervened by governments including departments of planning administration, cultural relics department, and real estate division, which form an approach combining governments' support, enterprises' operation, and citizens' participation

Source: author summarizes according to the following research:
Guan(2008),Zhang(2006), Hu(2005), Huang(2008),Li(2010),Li and Lu(2005),Luo,et al.,(2002), Wang(2012),Xiao(2007),Xu, et al.,(2006),Yao and Zhao(2009),Zhu and Zhu(2010).

6.2 Conservation process of Li-Nong neighborhoods in Shanghai

Li-Nong, as a traditional dwelling type[①], have been recognized as a heritage type that were once largely existing while have now been demolished in large scale. The conservation of Li-Nong has been closely involved in the urban redevelopment and renewal process of Shanghai, hence mostly causing complicated situation in conservation practices on the historical quarters.

---

① Li-Nong architectures in Shanghai are normally categorized in five types: the early stage of old-style Li-Nong, the latter stage of old-style Li-Nong New-style Li-Nong, Apartment Li-Nong , and Garden Li-Nong. Zhang, s. (2006). "Conservation Strategy of Urban Heritage In Shanghai (上海城市遗产的保护策略)." City planning review30(2): 49-54, Liu, S. (2011). The Research of Shanghai Longtang Modern Historical Changes and Cultural Values (近代上海弄堂演变及其文化价值研究), Zhejiang Normal University.

In 1843, Shanghai opened as one of the Treaty Ports and became the centre of trade because of its excellent location. However, the Shanghai Small Sword Party uprising in 1853 and the Taiping Rebellion in 1860 interrupted the prosperity and development. The old sections of the city were seriously damaged; therefore, people flooded on foreign concessions. From 1853 to 1855, the number of Chinese people on concessions increased from 500 to more than 20,000(Fan 2004). Taking advantage of the occasion, British speculators built simple rent houses which could be considered as the rudiment of Li-Nong houses (Ibid.). In 1869, real estate developers began to build two-floor houses of bricks and wood. Later on, compact planned townhouses with the layout of European cities were wildly adopted in order to improve plot ratio and maximum profits (Ibid.). Afterwards, with over a century of development and derivation, Li-Nong houses diverged into different types, such as apartment-style, Shikumen-style, and Garden-style, and so on. The thesis focuses on the conservation process of Li-Nong houses after 1949, which could be divided into 4 stages as follows.

6.2.1 Conservation process

1) Environment improvement from 1950s to 1970s

In early 1950s, Shanghai started the renewal of old residential areas. Due to the financial and political situation at that time, the renewal was mainly to improve outdoor environment of residential areas, while in late 1950s to early 1960s, the main point transformed to indoor environment in order to improve living conditions (Fan 2004). But in early 1960s to late 1970s, the renewal process was interrupted for a long time by Culture Revolution.

2) The top-down large scale urban redevelopment of Li-Nong area during 1980s

In early 1980s, with the main objective of solving residence shortage, Shanghai restarted another redevelopment, in which large number of Li-Nong houses were demolished (Fan 2004). According to Guan (2008) there were three main issues that compromised the conservation of Li-Nong houses in 1980, including capital source, executors, and ways of operation. Firstly, since raising fund is the key points in city renew, local governments employed many methods to raise money, however, it has not become market-oriented and governments' financial supports covered most of fund resource (Guan 2008). Secondly, the renewal at that period was mainly to reuse the Li-Nong houses and improve the indoor space. Therefore, the executors were only government and its subordinate real estate development companies (Guan 2008). Thirdly, because city renewal at that time was under the guide of planned economic system, most shabby houses were demolished and reconstructed and after the process, local people moved back to new houses which belonged to their employers which left them with no property rights (Guan 2008).

In the middle 1980s, more and more people recognized the significance of Li-Nong in Shanghai and pay attention to its unique architecture style; meanwhile, not only the Li-Nong's physical fabric was recognized to have significant value, its social cultural values have been gradually calling the attention of society (Fan 2004).

3) *The policy led inner city renewal in 1990s*
In 1990s, the three issues introduced above were changed with the development of market economy. According to Zhao et al, (Zhao, Bao et al. 1998), the reform of Chinese land system offered new capital source for urban construction. Renewing old city area through real estate development could sufficiently utilized the effect of differential rent of urban land; meanwhile attracting foreign capital to remit finical shortage of the reconstruction was also a commonly adopted method in 1990s (Guan 2008). However, in this period, government was still the main executors of urban renewals although the real estate developers were getting involved. In the case of Shanghai's city centre renewal, governments at all levels served as front-runners at all activities, including directing, orientating, coordinating, supervising, and implementing (Xu 2004). Regarding the ways to operate, the supply of land was offered by subscribing agreements, and the land market was not fully formed (Guan 2008). Governments made demands to companies and offered them policy and finance supports, which made it possible to demolish a large amount of old houses at a short time; however, local residents were excluded from the process (Guan 2008). As a consequence, the way of urban renewal at that time was to demolish old houses and relocated the residents.

4) *Real-estate initiated renewal and the rise of historic district conservation in 2000s*
In 2000s, real estate developments had become the main approach of land replacement and planning authorities gave more space to the developers in order to diversify the land use in downtown area (Xu 2004). Meanwhile, with the hope of raising fund from markets and the intention to support and regulate the developments of real estate, it was proposed that the mechanism of urban renewal in downtown should creatively be transformed from the simple government-oriented approach to a government-supported, companies-executed and public-involved one (Xu 2004).

However, similar with the case in Xi'an, the conservation of historical areas in Shanghai was also threatened by the repaid urbanization and large-scale of urban reconstruction. In order to respond, "the regulation for conservation of areas with historical and cultural features and excellent historical buildings (上海市历史文化风貌区和优秀历史建筑保护条例)" was passed in July, 2002 and put in practice with the force of law in January, 2003. Following this regulation, Shanghai government designated twelve areas with historical and cultural features and developed the detailed regulatory planning for them, through which it can be seen that government paid equal attention to urban conservation compared to urban development and it became one the most important parts of urban planning in this stage (Zhang 2006). Some details of this planning will be discussed in the next section.

6.2.2 Current difficulties to conserve Li–Nong houses

Although local governments and planning authorities have put quite a lot of efforts on the conservation and renew of Li-Nong houses, certain challenges exist and make them difficult to carry out smoothly.

- *Over-crowdedness and property rights*

Similar with the case study in Xi'an, with a long history and the huge increase of population in cities, historical buildings especially residential buildings are overloaded. Li-Nong houses originally built for one family now occupied by several families, and meanwhile some of them transformed from single residential function into multi-functions, commonly with commercial functions in the ground floors like convenient stores, restaurants, and coffee shops and offices and living space in the upper floors. Because of the reasons above and multiple alterations of ownership over time, it is difficult to define the property right of some historical buildings, which compromises the management and preservation of historical buildings.

- *Difficulty to relocate residents*

In Shanghai, Most Li-Nong houses locate in the centre of downtown area, which has a relative convenient surrounding environment in terms of accessibility to facilities (Yu 2010). Therefore, a large number of residents here are willing to remain and eager to improve their poor indoor living conditions, especially aged people who are not willing to change their living habits and leave their familiar neighbors, which cause the aging tendency of population (Ibid). Because quite a lot of Li-Nong houses are over-crowed, so it is necessary to relocate some residents for the purpose of reducing population density. Nevertheless, local residents' request for the improvement of living environment without relocations contrary to the mainstream approach adopted by governments which is mass demolition and relocation.

- *Structure damage*

Besides, most of the historical buildings and blocks were built in around 1920s with nearly 70 to 80 years of history. Despite having been repaired, the improper use resulted in damage to structure and common staleness of inner facilities (Zhang 2006). With long time overload operation, the main structures and appearances of the historical buildings have been damaged severely (Ibid). In some cases, the improper change of function also causes irreversible destructions.

- *Incompletion of the system*

Among all difficulties, the most challenging issue is the incompletion of urban conservation system. According to Zhang(2006), laws and regulations to protect historical buildings and neighborhoods are not enough and sometimes urban conservation may be compromised by the ambition for economic development and rapid urbanization. According to Yu (2010), under the guide of land leasehold system in Shanghai and with the purpose to maximize profits, many Li-Nong houses in Shanghai have been demolished to build high rise residences and the land-transferring fees became the main source for large scale urban constructions since 1992. More importantly, until now, local residents are still uninvolved in the process of urban conservation and their rights and interest are usually ignored to some extent (Zhang 2006).

6.2.3 The values of Shanghai Li-Nong houses

According to Li (2010), the value of Li-Nong houses in Shanghai depends on their identity at different periods of time, and their identity is decided by the behavioral agents and their attitudes toward Li-Nong houses. Table 6-2 summarizes the change of Li-Nong houses' value from 1980s to 2000, in which it can be seen that behavioral agents changed from governments in 1980s to academic communities in 1990s and to developers in 2000s. Due to their different interests and attitudes, Li-Nong houses were considered as urban habitats, and potentials for economic developments. Therefore, in 1980s, because there was no value at all was recognized by local governments, quite a lot of Li-Nong houses were demolished in order to make room for new developments and provide modern living space for residents. In 1990s, academic communities, especially architects and urban planners emphasized that urban history should be respected and Li-Nong houses with significant historical value were conserved. However, "insignificant" ones were demolished. In 2000s, the behavioral agents became private developers and their interest was economic benefits and developments. Consequently, adaptive reuse became the popular method of conservation, which has a characteristic of gentrification (Ibid).

The value of Li-Nong houses changes over time   Table 6-2

	1980s	1990s	2000s
Behavioral agent	Local governments	Academy communities	Developers
Dominate ideas and attitudes towards Li-Nong houses	Traditional buildings represent outdated	urban history should be respected	Economic development is the priority
Objectives of developments	Urbanization and providing modern living space for residents	Urban image	Economic benefit and development
Li-Nong houses' identity	Urban slams	Urban heritage	Potentials for economic development

	1980s	1990s	2000s
Li-Nong houses' value	No value	Historical value and architectural value	Economic value
Actions on Li-Nong houses	Demolished	Those with significant historical values were conserved, but others were demolished.	Adaptive reused

Source: author summarizes according to (Li 2010).

Li (2010) argued that this mechanism of identifying Li-Nong houses has a great defect due to the oneness of behavioral agent, and he advocated a diversified mechanism to identify the value of Li-Nong houses. Li-Nong house is a complicated complex of space and society and a number of stakeholders are involved (Ibid.). It was also emphasized in his research that the social value and the value for everyday life were ignored because local residents and their interests were not included (Ibid.).

6.3 Urban conservation in Tianzi Fang

6.3.1 The evolution of Tianzi Fang (pre-conservation)

Originated from a creative reuse of abandoned factories, the development path of Tianzi Fang reveals a journey from place making to place management through collective efforts, and along this journey the value of a living urban heritage are explored and reshaped into the public realm, where cohesion is finally achievable.

1) Background of Taikang Road

The so called "Tianzi Fang" includes the traditional Li-Nong (里弄) neighborhood in Taikang Road (Alley No. 248, No.274, No.210) and the neighborhood factories in Alley No.210. Taikang Road was designated as French concession area (Figure 6-2, 6-3, and 6-4) during 1920s, and after years of transformation, the architectural manifestations are the mixed land use of residence and industries (Figure 6-5 and 6-6), and the mixed types of Shikumen (石库门) residential architecture. From 1949 till the dawn of the open policy, Taikang Road experienced the change of property rights from private to collective, the redistribution and subdivision of residential buildings, and reconstruction of factories upon old factory sites. Like many traditional Li-Nong neighborhood, Taikang Road faced serious physical degradation since early 1990s.

Along with the deepening of the reform, Shanghai launched the industrial restructuring and relocation, the large scale redevelopment of old quarters, and the reform of governance in 1990. Accompanied with the development of Pudong New District in early 1990s, Shanghai adjusted

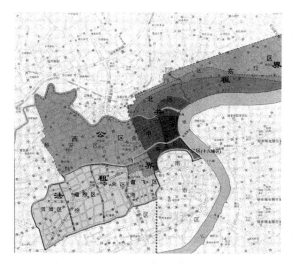

Figure 6-2　French concession area

Source: The Conservation Concept Plan of Taikang Road Historic District,Tongji University,2004.

Figure 6-3　Taikang Road historic area

Source: The Conservation Concept Plan of Taikang Road Historic District, Tongji University, 2004.

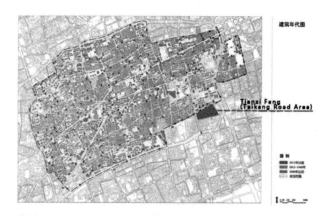

Figure 6-4　Taikang Road area

Source:Shanghai Historic District Conservation Plan——District of Hengshan Road—Fuxing Road, 2004, author edit.

Figure 6-5 Mix-used neighborhood, Li-Nong and neighborhood factory
Source : The Conservation Concept Plan of Taikang Road Historic District,Tongji University, 2004.

Figure 6-6 *Nongtang* styled fabric and diverse combination
Source: The Conservation Concept plan of Taikang Road Historic Drstrict, Tong Ji University, 2004.

its industry structure, which aims to develop service industry in central city, to relocate the manufactory industry, and at the same time to enhance the primary industry (Wu 2000). Many inner city factories faced the policy called "GUAN-TING-BING-ZHUAN"(关停并转).[①] In this process, around 700 factories in central city were relocated and 300 hm² of land became available for redevelopment (Yao 2008).

Meanwhile Shanghai launched a large scale inner city redevelopment. Firstly stimulated by the industrial restructuring (Guan 2008, Li 2010) as the abandoned industrial buildings became the

---
① A policy to close up factories, to suspend operation, to combine several factories into big one, and to transfer to the tertiary industry.

focus of the redevelopment (Ruan and Zhang 2004, Zhang and Chen 2010), inner city redevelopment in 1990s also included shanty residential areas. By 1997, 71 blocks covering an area of 3378,000 square meters located in Taikang Road of Luwan District were redeveloped (Yao 2008). Even though Taikang Road was not categorized as shantytowns, the shantytowns redevelopment examples are evident in Taikang Road's redevelopment efforts.

The last decade of 20th century also witnessed the institutional restructuring and the decentralization of power in Shanghai——A new administrative system composed of two levels of governments (municipal and district), and three levels of governance (municipal—district—sub-district). The decentralization of power and governance has empowered sub-district office with authority over population management, security management, economic development as well as the refunds of taxes and redevelopment income (Huang 2008). Accordingly, many rapid redevelopment projects are promoted by sub-districts.

It is under such social-economic and governance background that many old quarters in central Shanghai have experienced the rapid mass demolition and mass construction process. Tianzi Fang's development path however rooted in the same historic context, departures from the ordinary path and represents a creative development path of place making and place management. The development path of the so called Tianzi Fang area can be examined in three stages, namely the initial stage, the transitional stage and the target stage. The focus is on the decision making process of place making and place management, in which governance, development approaches, social cooperation, and collective action will be examined.

6.3.2 The development path of Tianzi Fang's conservation: *from place making to place management*

1) *The initial stage(1997-2002): from Taikang Road Art street to Shanghai SOHO*

After administration distribution, Taikang Road was under the administration of Dapuqiao sub-district in 1996. By 1997, due to the economic crisis in Asia, many redevelopment projects in Shanghai were either suspended or delayed. Led by Zheng Rongfa, Dapuqiao sub-district office revived the economy of this sub-district through a cultural economy approach.[①]

Taikang Road was not a random choice for art street development. A "Dumb-bell" plan (哑铃计划) was proposed which aimed to link two cultural and recreation facilities with Xujiahui Road and to develop Xujiahui Road as a cultural industrial destination (Zhang 2009). Zhan Jianjun, a former officer in charge of redevelopment and construction in Luwan District was invited

---

① *See the Blog of Zhang Jianjun, source: http://blog.sina.com.cn/s/blog_492698b10100d2na.html.*

as a consultant to provide expertise. He had suggested choosing a minor road rather than an arterial road in order to improve accessibility and attract people. This advice was accepted and an economic revitalization plan called "street economy"(街巷经济) was raised, which aimed to attract potential investors in Dapuqiao area.

Taikang Art Street (Figure 6-7) was implemented as a sub-district led Art street project, which was followed by a series of practice. Firstly, in order to improve the environment of Taikang Road, the street market was moved to an indoor market (Figure 6-8) renovating an abandoned factory (Zhang 2009, Zhu 2009). Secondly, two abandoned factories in Taikang Road were borrowed by the sub-district to accommodate cultural trade(Hu 2008).

Figure 6-7　Taikang Art Street
Source: Tongji University, Taikang Road Conservation fieldwork,2004.

Figure 6-8　Indoor market "Yilufa" which is renovated from Li-Nong neighborhood factory
Source: Tongji University , Taikang Road Conservation fieldwork,2004.

In 1998, a businessman, Wu Meisen was introduced to Zheng Rongfa as a business planner. Considering Wu's work experience as a cultural businessman and his connections with artists, the sub-district changed its plan to develop cultural economy and to rely on social resources for economic development.

Without increasing the rent, the sub-district sublet the two factories to Wu Meisen, who in turn, invited famous artists such as Chen Yifei, Huang Yongyu and Er Dongqiang to visit the factories in Taikang Road Alley No. 210. Attracted by the unique aroma of traditional Li-Nong neighborhood, they all supported the idea of artist factories (Figure 6-9). Chen Yifei identified the value of these Li-Nong factories as "Shanghai SOHO". Huang Yongyu renamed this alley with "Tianzi Fang" (田子坊). Following the famous artists, many artists crowded into this alley because of low rents. Li-Nong factories were renovated by artists into studios and galleries. In 2001, Tianzi Fang (Taikang Road No. 210) –Shanghai SOHO, was nominated by municipal government as one of the first 18 cultural industry hubs (Yao 2008).

Figure 6-9   Artist factory——an adaptive reuse of Li-Nong factories
source: Tongji University, Taikang Road Conservation fieldwork,2004

Given the condition of investment shortage from real estate due to the impact of Asian Economic Crisis, this development plan was supported by Luwan District government. Two organizations were established during the initial stage. A governmental organization and Taikang Art Street Committee was established to facilitate the implementation of the art street project. The committee was managed by a street office. In 2002, Taikang Road Artist Union, a non-governmental organization, chaired by Chen Yifei was founded, which communicate with Taikang Art Committee (Hu 2008) .

Under the context of the economic, social and institutional reform, the economic restructuring and industrial relocation provided the space for urban redevelopment, which was driven by the need for physical revitalization in inner city. And the administrative restructuring empowered the sub-districts; while at the same time urges them to compete with each others. Thus, it is understandable that priority was given to economic development. The Asian Economic Crisis slowed down the redevelopment projects, which was the mainstream approach for economic development in inner city. From "Art street economy" to "Shanghai SOHO", Dapuqiao Sub-district initiated an alternative approach for economic development, which did not rely on the municipal and district governments' financial allocation and the investment of real estate, but by cooperation with social resources, conducted a successful place making practice: improving the environment, creatively reusing the abandoned factories, and reducing the burden of government (Zhang 2009, Zhu 2009).

2) *The transitional stage (2002—2007): from place making to heritage conservation*
As the city gradually recovered from the impacts of the Asian Economic Crisis in early 2000s, inner city redevelopment were re-launched. The target of redevelopment in the second round include "Liang Ji Jiu Li"(两级旧里)[①], and the "Xin-Xin Li"(新新里) area (Figure 6-10) which covers Northern Taikang Road neighborhood. As can be seen from (Figure 6-11, 6-12 and 6-13), the dwelling condition in Taikang Road neighborhood is quite poor.

Figure 6-10　The demolished Li-Nong neighborhood Xin-Xin Li
Source: Tongji University, Taikang Road Conservation fieldwork.

---

① *According to the categories of housing in Shanghai, shikumen housing without toilets are included.*

Figure 6-11　Shared kitchen
Source: Ibid.

Figure 6-12　Dilapidated facilities
Source: Ibid.

Figure 6-13　Cramped living space
Source: Ibid.

 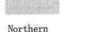

Figure 6-14 Xin-Xin Li Redevelopment Plan (2003) which includes northern and southern parts
Source: Google map, edited by author.

According to the Regulation Detail Plan of Luwan District, the redevelopment of Xin-Xin Li (Figure 6-14 and 6-15) area will be a significant project in Dapuqiao area.① In October 2003, the Regulation Detail Plan of Xin-Xin Li Area was approved. According to this plan, the neighborhood in Taikang Road Valley No. 210, 274 and 248 will be demolished and redeveloped for high-rise commercial and residential use.② A subsidiary company of Taiwan ASE Group Ltd, Shanghai Ding-Rong Real Estate Ltd acquired the land development right. In early 2003, a demolition notice was released by Luwan District (Zhang 2009). It is clear that in the transitional stage, the lock-in effect of the mainstream redevelopment policy favors a short-term redevelopment process due to the pressure of housing and economic development. Nevertheless the transitional stage saw the creation of an alternative development path, in which the values of Tianzi Fang have been reinterpreted and reconstructed by different social groups through a networking social capital heritage initiated conservation decision making process.

The process of decision making to conserve Tianzi Fang can be traced in two steps. In the first step, the focus is to protect the artist factories from demolition (Figure 6-16). Different social groups participated in the campaign to conserve the artist factories. The sub-district secretary of the party committee, Zheng Rongfa fought for Tianzi Fang in the administration system. Famous artists including Chen Yifei, Er Dongqiang and other influential artists appealed to the

---

① see *Xujiahui Road Development Plan,1994*.
② see the *Regulation Detail Plan of Xin-Xin Li Area,2003*.

Figure 6-15 Xin-Xin Li Redevelopment Project and Taikang Road Neighborhood in the north
source: Mr. Zheng Rongfa.

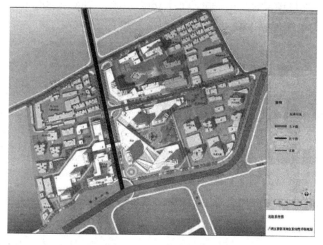

Figure 6-16 Detail Plan of Xin-Xin Li area 2003 and the conserved factories area (red marked buildings) in 2004
Source: Regulation Detail Plan of Xin-Xin Li Area 2003, edited by author.

government for the conservation (Zhang 2009). Renowned Scholars, specialists and research institutes were invited by Wu Meisen.[①] The value of Tianzi Fang was examined from aspects of architectural value, historic value, and urban living heritage value, social and economic value. Especially during 2002—2004, creative industry was widely discussed and promoted in important newspapers and media. In October 2004, the Regulation Detail Plan of Xin-Xin Li Area

---

[①] *Architectural Culture scholars like Zheng Shiling, conservation specialist like Ruan Yisan, creative industry researcher Li Wuwei,etc.*

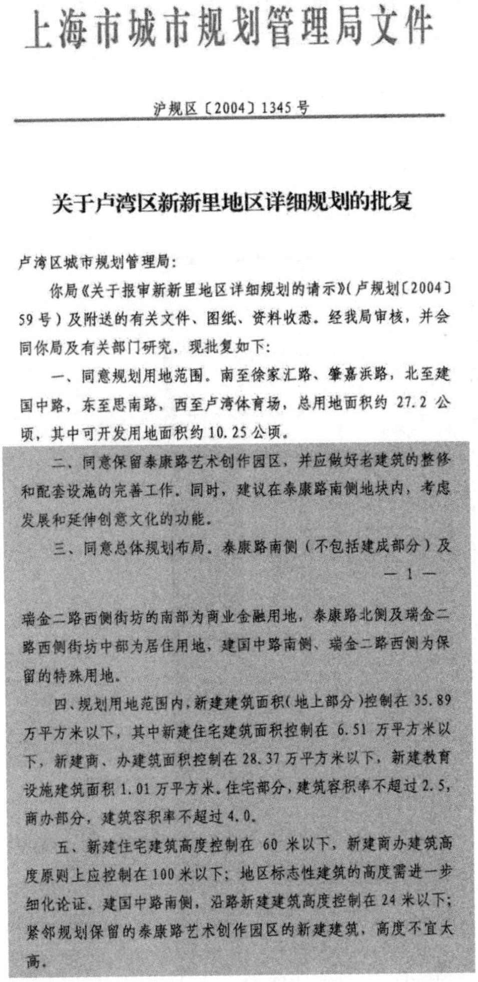

上海市城市规划管理局文件

沪规区〔2004〕1345号

关于卢湾区新新里地区详细规划的批复

卢湾区城市规划管理局：

你局《关于报审新新里地区详细规划的请示》(卢规划〔2004〕59号）及附送的有关文件、图纸、资料收悉。经我局审核，并会同你局及有关部门研究，现批复如下：

一、同意规划用地范围。南至徐家汇路、肇嘉浜路，北至建国中路，东至思南路，西至卢湾体育场，总用地面积约27.2公顷，其中可开发用地面积约10.25公顷。

二、同意保留泰康路艺术创作园区，并应做好老建筑的整修和配套设施的完善工作。同时，建议在泰康路南侧地块内，考虑发展和延伸创意文化的功能。

三、同意总体规划布局。泰康路南侧（不包括建成部分）及

— 1 —

瑞金二路西侧街坊的南部为商业金融用地，泰康路北侧及瑞金二路西侧街坊中部为居住用地，建国中路南侧、瑞金二路西侧为保留的特殊用地。

四、规划用地范围内，新建建筑面积（地上部分）控制在35.89万平方米以下，其中新建住宅建筑面积控制在6.51万平方米以下，新建商、办建筑面积控制在28.37万平方米以下，新建教育设施建筑面积1.01万平方米。住宅部分，建筑容积率不超过2.5，商办部分，建筑容积率不超过4.0。

五、新建住宅建筑高度控制在60米以下，新建商办建筑高度原则上应控制在100米以下；地区标志性建筑的高度需进一步细化论证。建国中路南侧，沿路新建建筑高度控制在24米以下；紧邻规划保留的泰康路艺术创作园区的新建建筑，高度不宜太高。

Figure 6-17: Document for conserving the factories in 1st step, 2004
Source:http://218.242.36.250/News_Show.aspx?id=8587&type=1, edited by author.

was adjusted by Luwan District. The artist factories in Taikang Road No. 210 were protected from demolition, while the residential area remains to be demolished (Figure 6-17).

In the second step, the focus is to protect the neighborhood in Taikang Road Alley No. 274, 248 and 210 from demolition. A grassroots movement played an important role in this step. The previous efforts to resist demolition has slowed down the implementation of the Xin-Xin Li redevelopment, at the same time, the increasing demand for space to accommodate cultural industry has initiated a bottom up activity——functional change of residential buildings into commercial premise——"居改非"(Figure 6-18), which is started by residents to lease out their house to artists and change residential buildings into commercial use (Zhang 2009, Zhu 2009). The agglomeration effect of art factories assured a considerable rental income for residents. And the residents started to get benefit from rental difference with adjacent area.① Zhou Xinliang took the initiative and many residents followed him. Compared to the one-off demolition compensation and the long-term rental income, most of the residents realized the value added by the creative hub "Tianzi Fang".②

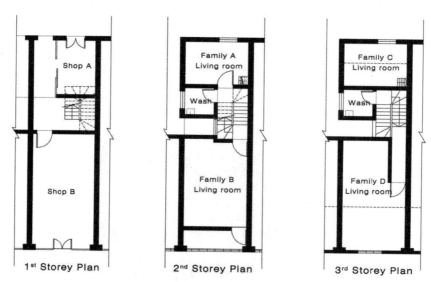

Figure 6-18　functional reuse of residential units in Tianzi Fang
Source: Pwee, 2011, p. 72.

A grass-roots organization was established as "Tianzi Fang Shikumen Owner Committee", with the aim to help residents rent their houses and manage the building. The Owner Committee self-financed to improve environment of "Er Jing Xiang"(二井巷) in order to attract tenants while petitioning to district government and municipal government for conservation of the residential area (Figure 6-19).

In 2007 Real Right Law of the People's Republic of China was released. Under article 76, matters regarding "rebuilding or any of its affiliated facilities", and matters regarding "the common ownership and the common management" shall be a common decision by owners (Jiang, Ma et al. 2009). According to the survey conducted in 2007, among total 671 households of the entire Tianzi Fang area, 363 households (54.67%) rented out their house for commercial and other

---

①　*According to Zhou Xinliang, in 2004, his retirement pension was only ¥339/month, which was barely enough. He had to rely on relative's help. When he rented out his 40sq.m room in 2004, the rental income was ¥3500/month. He spent 1,000 to rent a house nearby, and the rental difference 2,500 large improved his living quality. The rent for his 40sq.m room in 2012 was ¥15,000/month.*
②　*According to 2004 Shanghai demolition compensation standard, Taikang Road area is 7550Yuan/Square meters. As most of Tianzi fang household's living space is less than 30 Square meters, the compensation will be around ¥210,000/ household*

uses, and 77% households were willing to exchange through the soft approach. And only 97 (14.6%) households chose to continue living in Tianzi Fang, which means that conservation of the entire Tianzi Fang and the development of creative industry has been largely accepted by the residents.① With the expansion towards neighborhood, Tianzi Fang ranked Top 3 national creative industry community in 2006.

Figure 6-19　Functional reuse of Er Jing Alley

In 2004, Zheng Rongfa was transferred to the head of Luwan District Association of Science and Technology (Zhang 2009). In 2006, proposed by Wu Meisen, Tianzi Fang Intellectual Property Alliance was established (Figure 6-20, 6-21, and 6-22). By linking creative industry and the official Luwan District Association of science and technology, Zhong Rongfa successfully retained the discourse power and dedicated in the campaign of conserving Tianzi Fang.

Figure 6-20　House owners signed jointly to protect the Intellectual Property
Source: Mr. Zheng Rongfa.

---

① *Based on the record in Dec 2012, there are totally 426 shops, over 300 of which are renovated from residential building. Among 671 households, over 500 have rent out. Around 80 households live in Tianzi Fang, among which, only 20 hold negative feelings towards the soft redevelopment approach.*

Figure 6-21　Tenants signed jointly an agreement to protect the Intellectual Property
Source: Mr. Zheng Rongfa.

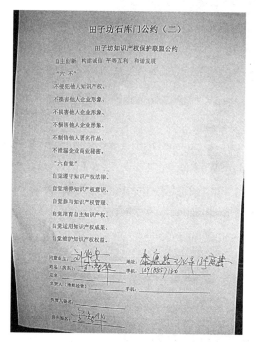

Figure 6-22　The agreement to protect to protect the Intellectual Property
Source: Mr. Zheng Rongfa.

Figure 6-23　Physical improvement in Tianzi Fang
Source: Mr. Zheng Rongfa.

After years of redevelopment with mass demolition and mass construction, there emerged a reflection on the rapid redevelopment approach, especially its impact on society and built environment (Ruan and Sun 2001, Ruan, Zhang et al. 2003). It is when other "JIU LI", namely traditional Li-Nong were demolished that the value of Tianzi Fang's survival as a living heritage have been fully recognized (Zhang 2009). The soft redevelopment approach in Tianzi Fang's received more and more recognition from both inside and outside the administration system. By 2007, the collective resistance against demolition has drawn the attention to Luwan District.

Under these changing circumstances, in 2007 Luwan District reconsidered the soft redevelopment approach and decided to conserve the entire area and promote the creative industry in Tianzi Fang.In addition physical improvement measures were conducted in Tianzi Fang (Figure 6-23). In Februry 2008 the Regulation Detail Plan of Xin-Xin Li Area was adjusted and the entire area was protected from being demolished. And in March 2008 "JU GAI FEI" in Tianzi Fang was approved as a pilot project to promote social cohesion (Zhu 2009) (Figure 6-24).

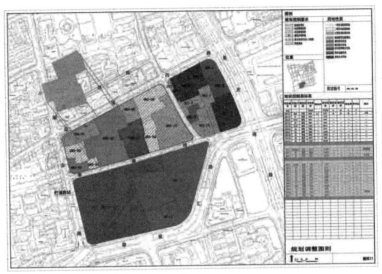

Figure 6-24 Land use of Tianzi Fang Neighborhood adjusted from residential to commercial-residential land and removed from redevelopment category
Source: Adjustment of Regulation Detail Plan of Huaihai Community (Unit C020101&C020102), 2008.

3) *The target stage (2008—): from heritage conservation to place management*
Shanghai Expo 2010 as an important event was linked with Tianzi Fang in Zheng's report to Luwan District 35th Executive Meeting in December 2007. In his report, Tianzi Fang will be targeted as a name card for Luwan District, a showcase to the world about Luwan's novel approach to promote urban regeneration and social cohesion through creative industry and heritage conservation. Tianzi Fang plays an important role in promoting tourism economy, creative economy, real estate market, as well as resolving unemployment and increasing revenue, which added a social value to the historic district.[①]

[①] Data shows that the tax revenues of Tianzi Fang in 2010 increased from 9.55 million Yuan in 2009 to 14.73million Yuan, with an increase of 64.8% on a year-on- year basis. From January 2011 to September 2011, the tax revenues amounted to 12.84 million Yuan, with an increase of 20% on a year-on- year basis. In 2012, the increase doubled. Source: http://baike.baidu.com/view/7185104.htm.

However, in the target stage, Luwan District was officially taken over in 2008. Tianzi Fang Management Committee was established on 2 April 2008, which includes a sub-district level office and a property management office. Luwan District adjusted policies to encourage the development of Tianzi Fang. Tax refund was given to Tianzi Fang Management Committee to improve environment and maintenance. Luwan District invested total 10 million Yuan to improve the physical condition of the entire area from 2008. Infrastructure and facilities such as automatic fire sprinkler system, sewage system, and electricity system, etc. are upgraded.[1] In 2010 Tianzi Fang was designated as a National AAA Tourist Spot and Urban Practices Area of Shanghai Expo 2010.[2]

The target stage reveals a top-down place management approach as decision making is limited to Tianzi Fang Management Committee whose members are all from Luwan District administration system. The Management Committee Office focuses on the implementation of creative industry, and the improvement of public administration and service. The property office took charge of security, building management, housing leasing and facilities management. According to the regulation, the land use change, the renovation of shops and subletting of houses should be reported to the Management Committee Office. But due to complicated property rights and subletting relationships, the management in Tianzi Fang was not satisfying.[3]

At the same time, many artists are fleeing from Tianzi Fang due to the rent increase. To meet the increasing rent, many art shops have been replaced by catering and hospitality industry. Numerous Bars, Cafes, restaurants, and souvenir shops have affected the everyday of residents and changed the art atmosphere. Tianzi Fang has been recently criticized for becoming a commercialized tourist spot.[4]

Horizontal dimensions of social resources continue to get involved in the management of Tianzi Fang. In 2009, Zheng Rongfa retired and established Shanghai Huaxia Creative Culture Research Institute on Taikang Road, a research association for promoting historic and cultural values of cities. Through organizing symposiums, exhibitions, workshops and other cultural activities and accompany visit, Tianzi Fang remains well connected with the institute and Zheng Rongfa.

---

[1] Last neighborhood physical upgrade was in 1980s.
[2] Report shows in recent years, tourist receipts of Tianzi Fang amounts to 200million per year. During Shanghai Expo 2010, the highest tourist receipts increase to 37,000 per day.
[3] According to the tourist satisfaction survey, Tianzi Fang ranked the bottom among all the AAA tourist spots in Shanghai. Source: http://collection.sina.com.cn/yjjj/20130328/0939108621.shtml
[4] According to Zhu Ronglin, from 2005 to 2010, creative industry in Tianzi Fang increased by 205%, while commercial business increased by 1231%, and the catering industry increased by 3400%. source: http://collection.sina.com.cn/yjjj/20130328/0939108621.shtml.

To protect the artist shops and creative industry and to facilitate the Management Committee, an organization of Tianzi Fang businessman, Tianzi Fang Trade Union, was established in March 2012. This union was chaired by Wu Meisen, who was excluded from the decision making group after the establishment of official Tianzi Fang Management Committee. Tianzi Fang Trade Union now has 35 members, most of which are from cultural and creative industry. In May 2013, the former Er Dongqiang Art Studio was renovated into Tianzi Fang Art Center and open to the public. According to Tianzi Fang Trade Union, the center is free for Artists to exhibit their works, and is also free for public to visit.

A bottom-up practice remains in this stage. Local residents who move out will occasionally gather through tea parties or festival dinners organized by Zhou Xinliang. They will receive an update with the information regarding the situation in Tianzai Fang. Zhou Xinliang and his group continue serve as property agents, communicating between residents and tenants.

Scholars and experts devoted their time in consulting and educating to secure a healthy business structure and a sustainable development of Tianzi Fang. Yu Hai, a professor of Fudan University has given several lectures to the house owners and analyzed the situation to prevent the unreasonable rent increase and further commercialization of Tianzi Fang.

In the target stage, art and creative industry are given priorities for Tianzi Fang's development as the value of the entire Tianzi Fang is targeted as a center of creative industry and arts. Collective efforts are made to promote the soft redevelopment approach. Though the top-down management Committee faces problems, it shares some common values with social groups towards the development of Tianzi Fang. There is no doubt when the bottom-up practices and horizontal intervention in keeping Tianzi Fang's value coexist with the management practice of the Committee, social cohesion could be achieved. And through collective decision making, with a well-connected network of social resources, the target stage of Tianzi Fang's development path towards social, economical and physical resilience in the long run will be assured.

6.3.3 A survey on local residents' perception towards Tianzi Fang's conservation

To grasp the entire image of Tianzi Fang residents' attitudes towards the development of Tianzi Fang, both remaining and moved-out residents were interviewed. There are total of 671 households in Tianzi Fang, according to the survey in December 2012, and over 500 households have rented out their "Tingzijian"(亭子间) among which, over 300 "Tingzijian" have been renovated into shops. The rest are mostly rented by employees working in the restaurants and shops in Tianzi Fang, and few are rented by migrant workers in the vegetable markets or nearby areas.

There are around 80 households remaining in Tianzi Fang. The following Tables (Table 6-3 to 6-7) will illustrate demographic profile, socio-economic profile, and change of dwelling in the last five to ten years, residents' perception on quality of life, and their attitudes towards future development.

Demographic profile Table 6-3

		Remaining residents N=41		Moved-out residents N=73		Total N=114	
Gender	Male	17	41.5%	34	46.6%	51	44.7%
	Female	24	58.5%	39	53.4%	63	55.3%
Age	18-29	2	4.9%	0	0.0%	2	1.8%
	30-39	1	2.4%	4	5.5%	5	4.4%
	40-49	3	7.3%	14	19.2%	17	14.9%
	50-59	8	19.5%	26	35.6%	34	29.8%
	60-69	13	31.7%	16	21.9%	29	25.4%
	≥70	14	34.1%	13	17.8%	27	23.7%

Socio-economic profile Table 6-4

		Staying residents N=41		Moving-out residents N=73		Total N=114	
Education level	None	18	43.9%	0	0.0%	18	15.8%
	Preliminary school and middle school	7	17.1%	33	45.2%	40	35.1%
	Technical secondary school and high school	9	22.0%	21	28.8%	30	26.3%
	Junior college Diploma	3	7.3%	6	8.2%	9	7.9%
	University and above	4	9.8%	13	17.8%	17	14.9%
Occupation	Student	2	4.9%	0	0.0%	2	1.8%
	Ordinary Worker	5	12.2%	4	5.5%	9	7.9%
	Technician	3	7.3%	7	9.6%	10	8.8%
	Manager	2	4.9%	3	4.1%	5	4.4%
	Self-employed	1	2.4%	15	20.5%	16	14.0%
	Retiree	28	68.3%	44	60.3%	72	63.2%
Length of residence	less than 3 years	3	7.3%	0	0.0%	3	2.6%
	3 – 10 years	2	4.9%	22	30.1%	24	21.1%
	10 - 20 years	4	9.8%	20	27.4%	24	21.1%
	over 20 years	32	78.0%	31	42.5%	63	55.3%
Ownership of flats in TZF	Private property	2	4.9%	1	1.4%	3	2.6%

		Staying residents N=41		Moving-out residents N=73		Total N=114	
	Old public housing	33	80.5%	72	98.6%	105	92.1%
	Rent	6	14.6%	0	0.0%	6	5.3%
Flat area	less than 30m^2	27	65.9%	52	71.2%	79	69.3%
	30-50m^2	11	26.8%	21	28.8%	32	28.1%
	50-80m^2	3	7.3%	0	0.0%	3	2.6%
	80-100m^2	0	0.0%	0	0.0%	0	0.0%
	over 100m^2	1	2.4%	0	0.0%	1	0.9%
Family size	1-2 people	4	9.8%	5	6.8%	9	7.9%
	3 people	9	22.0%	30	41.1%	39	34.2%
	4-5 people	21	51.2%	29	39.7%	50	43.9%
	5 and above	7	17.1%	9	12.3%	16	14.0%

As can be seen in Table 6-3, more than half (65.8%) of the respondents of remaining residents are over 60 years old, while among the moved out residents, respondents in their fifties to sixties are two major groups. Accordingly, occupation indicates that 63.2% of the respondents from both sides are retired senior citizens. With regard to education level, the remaining respondents' level of formal education is comparatively lower than the moved out respondents. As can be seen in Table 6-4, nearly half (43.9%) of remaining respondents failed to receive education; while more than half (54.8%) of the moved-out respondents received higher education than middle school. Data shows that most of the respondents live in Tianzi Fang for more than 10 years and the majority (92.1%) of respondents' apartments in Tianzi Fang belongs to the old public housing, which is directly managed by government. Most (69.3%) of the apartments in Tianzi Fang are less than 30m^2.

1) *Change in Tianzi Fang during last five to ten years*

Change of dwelling in recent 5-10 years Table 6-5

		Remaining residents N=41		Moving out residents N=73		Total N=114	
Leasing condition	Rent out	0	0.0%	73	100.0%	73	64.0%
	Intend to rent, in action	4	9.8%	0	0.0%	4	3.5%
	Want to rent but with difficulties	18	43.9%	0	0.0%	18	15.8%
	No intention	19	46.3%	0	0.0%	19	16.7%
Reasons of staying	Family reason-oldest old	12	29.3%				

		Remaining residents N=41		Moving out residents N=73		Total N=114	
Daily expenses in TZF	Family reason-joint property	7	17.1%				
	Upstairs-downstairs relationship	13	31.7%				
	Convenience	9	22.0%				
	Largely increased	17	41.5%				
	Increased	15	36.6%				
	Remain the same	6	14.6%				
	Reduced	3	7.3%				
Income	Largely improved	0	0.0%	49	67.1%		
	Improved	7	17.1%	22	30.1%		
	Remain the same	31	75.6%	2	2.7%		
	Worsened	3	7.3%	0	0.0%		
Current housing condition	Rent old public housing			7	9.6%		
	Rent a commercial flat			35	47.9%		
	Purchase a flat with mortgage			14	19.2%		
	Purchase a flat with full payment			17	23.3%		
Current dwelling quality	Largely improved			29	39.7%		
	Improved			43	58.9%		
	Remain the same			1	1.4%		
	Worsened			0	0.0%		

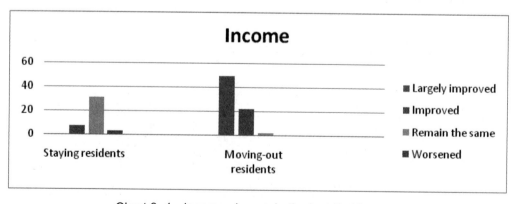

Chart 6-1 Income change in the last 5-10 years

How the development of Tianzi Fang in the last five to ten years has influenced local residents' lives is crucial in evaluating the value of heritage conservation. Table 3 describes the impacts of change on two groups. More than half (53.7%) of the remaining residents want to lease out their houses and move out. Currently, the reasons for remaining in resident can be understood as such; 1) more than a quarter of the respondents have an aged family member who refused to move out. 2) less than one fifth of the respondents share the joint property with other family members, which makes the segmented rental income unaffordable for living outside. 3) Around one third of the respondents live on the upper floors, given that flats in the ground floor remain residential, it is difficult to lease out their flats even though they are willing. 4) Around one fifth of the respondents choose to stay because the convenience of living in Tianzi Fang area, which is well connected with facilities and service.

With regard to the daily expenses, more than two third (78.1%) of the remaining respondents agree that the daily expenses have increased. Given that three quarters of the staying respondents' income remain unchanged or even devalued (Chart 6-1), it can be concluded that most of the remaining respondents have been affected by the increasing daily expenses. In contrast, the majority of the moved-out respondents agree that both their income (97.2%) and dwelling quality (98.6%) have been improved. It can be assumed that the rental income from Tianzi Fang flats make a better dwelling quality that is affordable to them.

2) The spatial transformation of Tianzi Fang
According to Hiroyuki Shinohara, the spatial transformation of Tianzi Fang from 2004 to 2009 can be summarized into the Figures 6-25. After 2008, Tianzi Fang continues to transform, with more and more residential space changed to commercial use and artist studios. A fieldwork in 2013 could indicate this further transformation (Figure 6-26).

In accompany with the physical transformation caused by functional change of Tianzi Fang, there coexists several processes of special practices such as 1) the creation of new artist space (Figure 6-27), 2) the continuous commercialization of Tianzi Fang (Figure 6-28), 3) the space of resistance (Figure 6-29), 4) the space of everyday life narrative (Figure 6-30), and so on and so forth. These features mingle with each other and coexist in the processes of reproduction Tianzi Fang.

A group of photos could reveal the physical changes in Tianzifang clearly, which indicate the transformation of a living Li-Nong neighborhood towards a tourism spot (Figure 6-31).

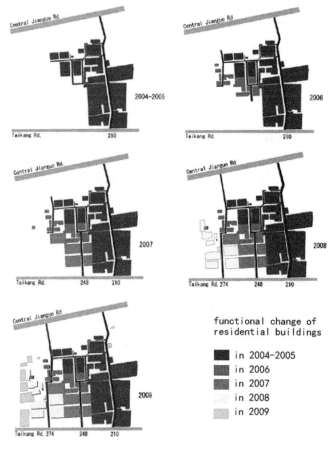

Figure 6-25　Spatial transformation of Tianzi Fang from 2004 to 2009
Source: Author edited according to (Shinohara 2009).

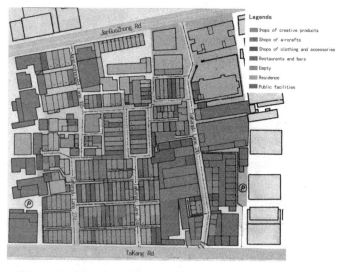

Figure 6-26　Functional layout of Tianzi Fang in 2013
Source: Fieldwork conducted by author.

Figure 6-27　Creative use of historic buildings

Figure 6-28　Commercialized Tianzi Fang

Figure 6-29　Space of unsatisfactory residents

Figure 6-30　Living narratives

Figure 6-31　Tianzi Fang in 2007
Source: Mr. Zheng Rongfa.

3) *Resident's attitudes towards Future Redevelopment*

Regarding the residents' attitude to the next stage of development in Tianzi Fang, different groups hold different opinions (Table 6-6). Nearly half (46.3%) of the remaining residents wish the entire area to be demolished and redeveloped, while more than half (67.1%) of the moved-out residents agree a small scale redevelopment which should replace only several deteriorated buildings. It is noticeable that both groups are aware of the current rental income and property value growth. As can be seen in Table 6-7, more than half (56.1%) of the remaining respondents are aware of the current rental income and property value growth, and the majority (97.3%) of the moved-out respondents insist that the current rental income and property value growth should be considered if there are any resettlement plans. In terms of resettlement means, more than half of respondents from both groups favor the flat area compensation means, but insist that resettlement should be in nearby area. With regard to their concerns of redevelopment, the staying respondents mainly worry about the adaptability and convenience of a new location. Nevertheless more than half (58.9%) of the residents are concerned about the loss of the continuous rental income and the growth of property value.

Resident's attitudes towards future redevelopment   Table 6-6

		Remaining residents		Moving-out residents		Total	
		41		73		114	
Scale of redevelopment	Entire TZF should be demolished	19	46.3%	5	6.8%	24	21.1%
	Most parts of TZF should be redeveloped	17	41.5%	17	23.3%	34	29.8%
	Small scale redevelop according to building quality	5	12.2%	49	67.1%	54	47.4%
	conserve entire TZF area	0	0.0%	2	2.7%	2	1.8%
Preferred resettlement Approach	Current flat area compensation means with resettlement in nearby area	23	56.1%	43	58.9%	66	57.9%
	Flat area prior to location	17	41.5%	9	12.3%	26	22.8%
	Consider current rental income and property value growth	22	53.7%	71	97.3%	93	81.6%
	Current money compensation means	11	26.8%	0	0.0%	11	9.6%
	Transaction between residents and customers	10	24.4%	5	6.8%	15	13.2%
	Residents negotiate and sell the entire TZF to a big company	7	17.1%	0	0.0%	7	6.1%
	Current leasing mode	2	4.9%	13	17.8%	15	13.2%
	Insist to stay	3	7.3%	0	0.0%	3	2.6%
Concerns of redevelopment	Close relationship with neighborhood	4	9.8%	0	0.0%	4	3.5%
	Difficult to adjust to new environment	9	22.0%	16	21.9%	25	21.9%
	Convenient facilities	17	41.5%	13	17.8%	30	26.3%
	Historical value	8	19.5%	1	1.4%	9	7.9%
	Continuous rental income and property value growth	3	7.3%	43	58.9%	46	40.4%

Residents' perception on the quality of life   Table 6-7

		Remaining residents N=41		Moving out residents N=73		Total N=114	
Income	Very satisfactory	0	0.0%	13	17.8%	13	11.4%
	Satisfactory	0	0.0%	43	58.9%	43	37.7%
	No objection	16	39.0%	11	15.1%	27	23.7%
	Unsatisfactory	11	26.8%	6	8.2%	17	14.9%
	very unsatisfactory	15	36.6%	0	0.0%	15	13.2%
Current dwelling condition	Very satisfactory	1	2.4%	3	4.1%	4	3.5%
	Satisfactory	13	31.7%	39	53.4%	52	45.6%
	No objection	5	12.2%	23	31.5%	28	24.6%
	Unsatisfactory	4	9.8%	9	12.3%	13	11.4%
	very unsatisfactory	18	43.9%	0	0.0%	18	15.8%
Living standards	Affluent	0	0.0%	0	0.0%	0	0.0%
	Well-to-do	3	7.3%	47	64.4%	50	43.9%
	No objection	21	51.2%	24	32.9%	45	39.5%
	Poor	17	41.5%	2	2.7%	19	16.7%
As TZF residents, you feel:	Very proud	4	9.8%	13	17.8%	17	14.9%
	Proud	9	22.0%	31	42.5%	40	35.1%
	Not bad	13	31.7%	25	34.2%	38	33.3%
	Inferior	15	36.6%	4	5.5%	19	16.7%
Quality of life	Improved	12	29.3%	62	84.9%	74	64.9%
	Worsened	25	61.0%	7	9.6%	32	28.1%
	Remained the same	4	9.8%	4	5.5%	8	7.0%

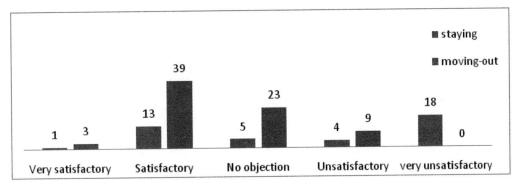

Chart 6-2 Current dwelling condition

Chart 6-2 reveals respondents' perception of their quality of life. Generally speaking, there exists a clear gap between two groups in terms of their perception on quality of life. More than half (76.7%) of them are satisfied with their current income. Only a few (12.3%) feel their current dwelling condition unsatisfactory. More than half (64.4%) of them agree that they are financially well-to-do, and the majority (84.9%) of them agree that the living quality has been improved. On the contrary, for the remaining respondents, more than half (63.4%) of them are unsatisfied with their current income. Only around one third (34.1%) are satisfied with their current dwelling condition (Chart 6-2). Even though more than half (58.8%) of them have no objections to their living standards, it is still noticeable that more than two fifth (41.5%) of the respondents feel they are poor. In the end, even though the government has upgraded the physical environment in Tianzi Fang , more than half (61.0%) of the remaining residents think their living quality has deteriorated, which could be explained by the disturbance of commercial activity and the increasing daily expenses. In addition, how residents currently feel about Tianzi Fang is very important to trace their attitudes towards their quality of life. More than half (60.3%) of the moved-out respondents feel proud about Tianzi Fang, and only a few (5.5%) feel inferior about being residents in Tianzi Fang. Data shows that the staying respondents hold complicated attitudes. Around one third (36.6%) of them feel inferior about being residents in Tianzi Fang. Nevertheless, nearly one third (31.8%) of them are beginning to feel proud of Tianzi Fang. Even though the remaining residents have not benefited from Tianzi Fang as much as the moved-out residents, the development path of Tianzi Fang which brings alternative opportunities to neighborhoods has changed the residents' original negative viewpoints, which saw the entire residential area only as "Liang-Ji-Jiu-Li" (两级旧里) , a desperate place waiting to be demolished.

Chapter 7   Data Analysis

This chapter aims to describe the data collected from social capital survey and social network analysis, and to discuss the results of data analysis in order to test hypothesis.

7.1   Community social capital based social cohesion analysis

7.1.1   Correlation analysis of variables within social cohesion indicator

The following tables present the correlation of the six variables to each other within each social cohesion cognitive indicator:

1) *Hui Fang, Xi'an*

Correlations analysis of variables of social cohesion indicators in Hui Fang, Xi'an   Table 7-1

		R1	R2	R3	R4	R5	R6
	Correlations analysis of variables of social cohesion indicator——Benevolence (仁)						
R1	Pearson Correlation	1	-.669**	.843**	.601**	.720**	.591**
	Sig. (2-tailed)		.000	.000	.000	.000	.000
	N	57	57	57	57	57	57
R2	Pearson Correlation	-.669**	1	-.669**	-.466**	-.614**	-.445**
	Sig. (2-tailed)	.000		.000	.000	.000	.001
	N	57	57	57	57	57	57
R3	Pearson Correlation	.843**	-.669**	1	.611**	.756**	.559**
	Sig. (2-tailed)	.000	.000		.000	.000	.000
	N	57	57	57	57	57	57
R4	Pearson Correlation	.601**	-.466**	.611**	1	.647**	.493**
	Sig. (2-tailed)	.000	.000	.000		.000	.000
	N	57	57	57	57	57	57
R5	Pearson Correlation	.720**	-.614**	.756**	.647**	1	.661**
	Sig. (2-tailed)	.000	.000	.000	.000		.000
	N	57	57	57	57	57	57
R6	Pearson Correlation	.591**	-.445**	.559**	.493**	.661**	1
	Sig. (2-tailed)	.000	.001	.000	.000	.000	
	N	57	57	57	57	57	57

** Correlation is significant at the 0.01 level (2-tailed).

Correlations analysis of variables of social cohesion indicator——Righteousness（义）

		Y1	Y2	Y3	Y4	Y5	Y6
Y1	Pearson Correlation	1	.377**	.593**	.587**	-.401**	.554**
	Sig. (2-tailed)		.004	.000	.000	.002	.000
	N	57	57	57	57	57	57
Y2	Pearson Correlation	.377**	1	.420**	.393**	.033	.590**
	Sig. (2-tailed)	.004		.001	.002	.807	.000
	N	57	57	57	57	57	57
Y3	Pearson Correlation	.593**	.420**	1	.738**	-.291*	.745**
	Sig. (2-tailed)	.000	.001		.000	.028	.000
	N	57	57	57	57	57	57
Y4	Pearson Correlation	.587**	.393**	.738**	1	-.427**	.579**
	Sig. (2-tailed)	.000	.002	.000		.001	.000
	N	57	57	57	57	57	57
Y5	Pearson Correlation	-.401**	.033	-.291*	-.427**	1	-.218
	Sig. (2-tailed)	.002	.807	.028	.001		.104
	N	57	57	57	57	57	57
Y6	Pearson Correlation	.554**	.590**	.745**	.579**	-.218	1
	Sig. (2-tailed)	.000	.000	.000	.000	.104	
	N	57	57	57	57	57	57

** Correlation is significant at the 0.01 level (2-tailed).

* Correlation is significant at the 0.05 level (2-tailed).

Correlations analysis of variables of social cohesion indicator——Propriety（礼）

		L1	L2	L3	L4	L5	L6
L1	Pearson Correlation	1	.813**	.730**	.776**	-.703**	.422**
	Sig. (2-tailed)		.000	.000	.000	.000	.001
	N	57	57	57	57	57	57
L2	Pearson Correlation	.813**	1	.792**	.798**	-.696**	.476**
	Sig. (2-tailed)	.000		.000	.000	.000	.000
	N	57	57	57	57	57	57
L3	Pearson Correlation	.730**	.792**	1	.874**	-.739**	.420**
	Sig. (2-tailed)	.000	.000		.000	.000	.001
	N	57	57	57	57	57	57
L4	Pearson Correlation	.776**	.798**	.874**	1	-.846**	.423**
	Sig. (2-tailed)	.000	.000	.000		.000	.001
	N	57	57	57	57	57	57
L5	Pearson Correlation	-.703**	-.696**	-.739**	-.846**	1	-.356**

							.007
L5	Sig. (2-tailed)	.000	.000	.000	.000		.007
	N	57	57	57	57	57	57
L6	Pearson Correlation	.422**	.476**	.420**	.423**	-.356**	1
	Sig. (2-tailed)	.001	.000	.001	.001	.007	
	N	57	57	57	57	57	57

** Correlation is significant at the 0.01 level (2-tailed).

Correlations analysis of variables of social cohesion indicator——Wisdom（智）

		Z1	Z2	Z3	Z4	Z5	Z6
Z1	Pearson Correlation	1	.318*	.308*	.435**	.310*	.339**
	Sig. (2-tailed)		.016	.020	.001	.019	.010
	N	57	57	57	57	57	57
Z2	Pearson Correlation	.318*	1	.579**	.453**	.666**	.429**
	Sig. (2-tailed)	.016		.000	.000	.000	.001
	N	57	57	57	57	57	57
Z3	Pearson Correlation	.308*	.579**	1	.545**	.524**	.329*
	Sig. (2-tailed)	.020	.000		.000	.000	.013
	N	57	57	57	57	57	57
Z4	Pearson Correlation	.435**	.453**	.545**	1	.436**	.272*
	Sig. (2-tailed)	.001	.000	.000		.001	.041
	N	57	57	57	57	57	57
Z5	Pearson Correlation	.310*	.666**	.524**	.436**	1	.378**
	Sig. (2-tailed)	.019	.000	.000	.001		.004
	N	57	57	57	57	57	57
Z6	Pearson Correlation	.339**	.429**	.329*	.272*	.378**	1
	Sig. (2-tailed)	.010	.001	.013	.041	.004	
	N	57	57	57	57	57	57

* Correlation is significant at the 0.05 level (2-tailed).

** Correlation is significant at the 0.01 level (2-tailed).

Correlations analysis of variables of social cohesion indicator——Faithfulness（信）

		X1	X2	X3	X4	X5	X6
X1	Pearson Correlation	1	.741**	.730**	.685**	-.650**	.607**
	Sig. (2-tailed)		.000	.000	.000	.000	.000
	N	57	57	57	57	57	57
X2	Pearson Correlation	.741**	1	.681**	.560**	-.758**	.709**
	Sig. (2-tailed)	.000		.000	.000	.000	.000
	N	57	57	57	57	57	57
X3	Pearson Correlation	.730**	.681**	1	.626**	-.600**	.507**
	Sig. (2-tailed)	.000	.000		.000	.000	.000

		N	57	57	57	57	57	57
X4	Pearson Correlation		.685**	.560**	.626**	1	-.515**	.504**
	Sig. (2-tailed)		.000	.000	.000		.000	.000
	N		57	57	57	57	57	57
X5	Pearson Correlation		-.650**	-.758**	-.600**	-.515**	1	-.700**
	Sig. (2-tailed)		.000	.000	.000	.000		.000
	N		57	57	57	57	57	57
X6	Pearson Correlation		.607**	.709**	.507**	.504**	-.700**	1
	Sig. (2-tailed)		.000	.000	.000	.000	.000	
	N		57	57	57	57	57	57

** Correlation is significant at the 0.01 level (2-tailed).

In summary, In the case of Hui Fang, the above tables (Table 7-1) illustrate that there commonly exist significant correlations among the six variables of each social cohesion indicators. In addition, it could also verify the validity of these variables of each cognitive social cohesion indicator.

2) *Tianzi Fang, Shanghai*

Correlations analysis of variables of social cohesion indicators in Tianzi Fang, Shanghai   Table 7-2

Correlations analysis of variables of social cohesion indicator——Benevolence (仁)							
		R1	R2	R3	R4	R5	R6
R1	Pearson Correlation	1	-.650**	.625**	.599**	.513**	.538**
	Sig. (2-tailed)		.000	.000	.000	.000	.000
	N	43	43	43	43	43	43
R2	Pearson Correlation	-.650**	1	-.579**	-.429**	-.390**	-.309*
	Sig. (2-tailed)	.000		.000	.004	.010	.043
	N	43	43	43	43	43	43
R3	Pearson Correlation	.625**	-.579**	1	.845**	.637**	.548**
	Sig. (2-tailed)	.000	.000		.000	.000	.000
	N	43	43	43	43	43	43
R4	Pearson Correlation	.599**	-.429**	.845**	1	.850**	.743**
	Sig. (2-tailed)	.000	.004	.000		.000	.000
	N	43	43	43	43	43	43
R5	Pearson Correlation	.513**	-.390**	.637**	.850**	1	.828**
	Sig. (2-tailed)	.000	.010	.000	.000		.000
	N	43	43	43	43	43	43

			.538**	-.309*	.548**	.743**	.828**	1
R6	Pearson Correlation							
	Sig. (2-tailed)		.000	.043	.000	.000	.000	
	N		43	43	43	43	43	43

** Correlation is significant at the 0.01 level (2-tailed).

* Correlation is significant at the 0.05 level (2-tailed).

Correlations analysis of variables of social cohesion indicator——Righteousness（义）

		Y1	Y2	Y3	Y4	Y5	Y6
Y1	Pearson Correlation	1	.665**	.775**	.567**	-.680**	.356*
	Sig. (2-tailed)		.000	.000	.000	.000	.019
	N	43	43	43	43	43	43
Y2	Pearson Correlation	.665**	1	.726**	.706**	-.618**	.433**
	Sig. (2-tailed)	.000		.000	.000	.000	.004
	N	43	43	43	43	43	43
Y3	Pearson Correlation	.775**	.726**	1	.779**	-.815**	.530**
	Sig. (2-tailed)	.000	.000		.000	.000	.000
	N	43	43	43	43	43	43
Y4	Pearson Correlation	.567**	.706**	.779**	1	-.719**	.409**
	Sig. (2-tailed)	.000	.000	.000		.000	.007
	N	43	43	43	43	43	43
Y5	Pearson Correlation	-.680**	-.618**	-.815**	-.719**	1	-.502**
	Sig. (2-tailed)	.000	.000	.000	.000		.001
	N	43	43	43	43	43	43
Y6	Pearson Correlation	.356*	.433**	.530**	.409**	-.502**	1
	Sig. (2-tailed)	.019	.004	.000	.007	.001	
	N	43	43	43	43	43	43

** Correlation is significant at the 0.01 level (2-tailed).

* Correlation is significant at the 0.05 level (2-tailed).

Correlations analysis of variables of social cohesion indicator——Propriety（礼）

		L1	L2	L3	L4	L5	L6
L1	Pearson Correlation	1	.650**	.556**	.517**	-.414**	.503**
	Sig. (2-tailed)		.000	.000	.000	.006	.001
	N	43	43	43	43	43	43
L2	Pearson Correlation	.650**	1	.732**	.714**	-.629**	.578**
	Sig. (2-tailed)	.000		.000	.000	.000	.000
	N	43	43	43	43	43	43
L3	Pearson Correlation	.556**	.732**	1	.742**	-.611**	.693**
	Sig. (2-tailed)	.000	.000		.000	.000	.000
	N	43	43	43	43	43	43

L4	Pearson Correlation	.517**	.714**	.742**	1	-.720**	.751**
	Sig. (2-tailed)	.000	.000	.000		.000	.000
	N	43	43	43	43	43	43
L5	Pearson Correlation	-.414**	-.629**	-.611**	-.720**	1	-.656**
	Sig. (2-tailed)	.006	.000	.000	.000		.000
	N	43	43	43	43	43	43
L6	Pearson Correlation	.503**	.578**	.693**	.751**	-.656**	1
	Sig. (2-tailed)	.001	.000	.000	.000	.000	
	N	43	43	43	43	43	43

** Correlation is significant at the 0.01 level (2-tailed).

Correlations analysis of variables of social cohesion indicator——Wisdom（智）

		Z1	Z2	Z3	Z4	Z5	Z6
Z1	Pearson Correlation	1	.686**	.702**	.652**	.660**	.643**
	Sig. (2-tailed)		.000	.000	.000	.000	.000
	N		43	43	43	43	43
Z2	Pearson Correlation	.686**	1	.779**	.636**	.725**	.601**
	Sig. (2-tailed)	.000		.000	.000	.000	.000
	N	43		43	43	43	43
Z3	Pearson Correlation	.702**	.779**	1	.716**	.737**	.619**
	Sig. (2-tailed)	.000	.000		.000	.000	.000
	N	43	43		43	43	43
Z4	Pearson Correlation	.652**	.636**	.716**	1	.638**	.673**
	Sig. (2-tailed)	.000	.000	.000		.000	.000
	N	43	43	43		43	43
Z5	Pearson Correlation	.660**	.725**	.737**	.638**	1	.601**
	Sig. (2-tailed)	.000	.000	.000	.000		.000
	N	43	43	43	43		43
Z6	Pearson Correlation	.643**	.601**	.619**	.673**	.601**	1
	Sig. (2-tailed)	.000	.000	.000	.000	.000	
	N	43	43	43	43	43	

** Correlation is significant at the 0.01 level (2-tailed).

Correlations analysis of variables of social cohesion indicator——Faithfulness（信）

		X1	X2	X3	X4	X5	X6
X1	Pearson Correlation	1	.785**	.660**	.746**	-.664**	.424**
	Sig. (2-tailed)		.000	.000	.000	.000	.005
	N		43	43	43	43	43
X2	Pearson Correlation	.785**	1	.733**	.841**	-.684**	.330*
	Sig. (2-tailed)	.000		.000	.000	.000	.031

	N	43	43	43	43	43	43
	Pearson Correlation	.660**	.733**	1	.816**	-.747**	.415**
X3	Sig. (2-tailed)	.000	.000		.000	.000	.006
	N	43	43	43	43	43	43
	Pearson Correlation	.746**	.841**	.816**	1	-.777**	.459**
X4	Sig. (2-tailed)	.000	.000	.000		.000	.002
	N	43	43	43	43	43	43
	Pearson Correlation	-.664**	-.684**	-.747**	-.777**	1	-.462**
X5	Sig. (2-tailed)	.000	.000	.000	.000		.002
	N	43	43	43	43	43	43
	Pearson Correlation	.424**	.330*	.415**	.459**	-.462**	1
X6	Sig. (2-tailed)	.005	.031	.006	.002	.002	
	N	43	43	43	43	43	43

** Correlation is significant at the 0.01 level (2-tailed).

* Correlation is significant at the 0.05 level (2-tailed).

In summary, in the case of Tianzi Fang, the above tables (Table 7-2) illustrate that there commonly exist significant correlations among the six variables of each social cohesion indicators. In addition, it could also verify the validity of these variables of each cognitive social cohesion indicator.

7.1.2 Data extraction——factor analysis

As explained in methodology, data extraction is applied to simplify the social cohesion indicators. Given that the 4 structural social capital indicators ($S1$-$S4$)) are simple indicators, 5 cognitive social capital indicators (R-Y-L-Z-X) are simplified through the extraction method of Principle Component Analysis (PCA). The results are summarized in the following part (Appendix A):

1) *Hui Fang, Xi'an*

Factor analysis of social cohesion indicator——Benevolence（仁）in Hui Fang

As shown in the table (Appendix A-1), after data extraction, the principle component could explain nearly 69% of the variance——Benevolence, and represents the information of variables respectively as followings : 89.6% of R1, 77.5% of R2, 90% of R3, 96.2% of R4, 88.8% of R5, and 74.6% of R6.

Factor analysis of social cohesion indicator——Righteousness（义）in Hui Fang

As shown in the above table (Appendix A-2), after data extraction, two components are identified as the principle components, which could explain nearly 73% of the variance——Righteousnessin total, and can carry the most of the information of the previous six variables.

Factor analysis of social cohesion indicator——Propriety (礼) in Hui Fang

As shown in the above table (Appendix A-3), after data extraction, the principle component could explain nearly 72.9% of the variance——Propriety, and represents the information of variables respectively as followings : 88.2% of L1, 90.6% ofL2, 90.7% of L3, 94% of L4, 86.5% of L5, and 56.1% of L6.

Factor analysis of social cohesion indicator——Wisdom (智) in Hui Fang

As shown in the above table (Appendix A-4), after data extraction, the principle component could explain nearly 52.4% of the variance——Wisdom, and represents the information of variables respectively as followings : 59.2% of Z1, 81.8% of Z2, 77.7% of Z3, 72.8% of Z4, 78.6% of Z5, and 60.8% of Z6.

Factor analysis of social cohesion indicator——Faithfulness (信) in Hui Fang

As shown in the above table (Appendix A-5), after data extraction, the principle component could explain nearly 70% of the variance——Faithfulness, and represents the information of variables respectively as followings : 88.2% of X1, 89.2% of X2, 82.6% of X3, 76.8% of X4, 84.4% of X5, and 80.1% of X6.

2) *Tianzi Fang, Shanghai*

Factor analysis of social cohesion indicator——Benevolence (仁) in Tianzi Fang

As shown in the above table (Appendix A-6), after data extraction, the principle component could explain nearly 67.7% of the variance——Benevolence, and represents the information of variables respectively as followings : 78.4% of R1, 65.2% of R2, 86.5% of R3, 92.3% of R4, 87% of R5, and 81.7% of R6.

Factor analysis of social cohesion indicator——Righteousness (义) in Tianzi Fang

As shown in the above table (Appendix A-7), after data extraction, the principle component could explain nearly 69.1% of the variance——Righteousness, and represents the information of variables respectively as followings: 82.3% of Y1, 84% of Y2, 93.9% of Y3, 85.1% of Y4, 87.9% of Y5, and 61.8% of Y6.

Factor analysis of social cohesion indicator——Propriety (礼) in Tianzi Fang

As shown in the above table (Appendix A-8), after data extraction, the principle component could explain nearly 69.58% of the variance——Propriety, and represents the information of variables respectively as followings : 71.4% of L1, 86.2% ofL2, 87.2% of L3, 89.6% of L4, 80.8% of L5, and 84% of L6.

Factor analysis of social cohesion indicator——Wisdom（智）in Tianzi Fang
As shown in the above table (Appendix A-9), after data extraction, the principle component could explain nearly 72.67% of the variance——Wisdom, and represents the information of variables respectively as followings : 84.9% of Z1, 86.8% of Z2, 89.4% of Z3, 84.3% of Z4, 85.4% of Z5, and 80.4% of Z6.

Factor analysis of social cohesion indicator——Faithfulness（信）in Tianzi Fang
As shown in the above table (Appendix A-10), after data extraction, the principle component could explain nearly 70.9% of the variance——Faithfulness, and represents the information of variables respectively as followings : 85.9% of X1, 88.8% of X2, 88.1% of X3, 93.5% of X4, 86.8% of X5, and 56.9% of X6.

7.1.3　Testing hypothesis 1 and hypothesis 2

After factor analysis, a new social cohesion construct which includes nine scales is developed to test the hypotheses. Five scales belong to cognitive social cohesion——"Benevolence-Righteousness-Propriety-Wisdom-Faithfulness(仁－义－礼－智－信)", and four belong to structural social capital——"Network integration, Resource network, Participation of formal and informal organizations". Each scale has only one principle component representing most of the variance. A correlation analysis is conducted between the new social cohesion construct and the Conservation Attitudes among local residents, and the results is summarized in the followings tables (Table 7-3, Table 7-4).

Correlations analysis of social cohesion constructs and resident's attitudes towards conservation in Hui Fang, Xi'an　　Table 7-3

	R(仁)	Y(义)	L(礼)	Z(智)	X(信)	S1	S2	S3	S4	CA
R Pearson Correlation (仁)Sig. (2-tailed) N	1 57	.526** .000 57	.579** .000 57	.596** .000 57	.785** .000 57	-.501** .000 57	-.193 .150 57	.573** .000 57	-.115 .396 57	.311* .018 57
Y Pearson Correlation (义)Sig. (2-tailed) N	.526** .000 57	1 57	.817** .000 57	.609** .000 57	.377** .004 57	-.431** .001 57	-.092 .494 57	.325* .014 57	.040 .767 57	.544** .000 57
L Pearson Correlation (礼)Sig. (2-tailed) N	.579** .000 57	.817** .000 57	1 57	.642** .000 57	.501** .000 57	-.535** .000 57	-.133 .323 57	.370** .005 57	.008 .953 57	.649** .000 57

	R(仁)	Y(义)	L(礼)	Z(智)	X(信)	S1	S2	S3	S4	CA
Z Pearson Correlation (智) Sig. (2-tailed) N	.596** .000 57	.609** .000 57	.642** .000 57	1 57	.454** .000 57	-.636** .000 57	-.257 .054 57	.250 .061 57	.161 .232 57	.334* .011 57
X Pearson Correlation (信) Sig. (2-tailed) N	.785** .000 57	.377** .004 57	.501** .000 57	.454** .000 57	1 57	-.468** .000 57	-.240 .072 57	.594** .000 57	.069 .611 57	.213 .112 57
S1 Pearson Correlation Sig. (2-tailed) N	-.501** .000 57	-.431** .001 57	-.535** .000 57	-.636** .000 57	-.468** .000 57	1 57	.586** .000 57	-.609** .000 57	-.085 .531 57	-.232 .082 57
S2 Pearson Correlation Sig. (2-tailed) N	-.193 .150 57	-.092 .494 57	-.133 .323 57	-.257 .054 57	-.240 .072 57	.586** .000 57	1 57	-.401** .002 57	-.125 .354 57	.012 .930 57
S3 Pearson Correlation Sig. (2-tailed) N	.573** .000 57	.325* .014 57	.370** .005 57	.250 .061 57	.594** .000 57	-.609** .000 57	-.401** .002 57	1 57	.112 .407 57	.158 .240 57
S4 Pearson Correlation Sig. (2-tailed) N	-.115 .396 57	.040 .767 57	.008 .953 57	.161 .232 57	.069 .611 57	-.085 .531 57	-.125 .354 57	.112 .407 57	1 57	.004 .979 57
CA Pearson Correlation Sig. (2-tailed) N	.311* .018 57	.544** .000 57	.649** .000 57	.334* .011 57	.213 .112 57	-.232 .082 57	.012 .930 57	.158 .240 57	.004 .979 57	1 57

** Correlation is significant at the 0.01 level (2-tailed).

* Correlation is significant at the 0.05 level (2-tailed).

Correlations analysis of social cohesion constructs and resident's attitudes towards conservation in Tianzi Fang, Shanghai Table 7-4

	R(仁)	Y(义)	L(礼)	Z(智)	X(信)	S1	S2	S3	S4	CA
R Pearson Correlation (仁) Sig. (2-tailed) N	1 43	.722** .000 43	.780** .000 43	.518** .000 43	.357* .019 43	-.052 .741 43	.226 .146 43	.153 .328 43	.357* .019 43	.690** .000 43
Y Pearson Correlation (义) Sig. (2-tailed) N	.722** .000 43	1 43	.730** .000 43	.482** .001 43	.317* .038 43	-.145 .354 43	.139 .376 43	.173 .267 43	.317* .038 43	.727** .000 43
L Pearson Correlation (礼) Sig. (2-tailed) N	.780** .000 43	.730** .000 43	1 43	.676** .000 43	.473** .001 43	-.255 .098 43	.104 .508 43	.248 .108 43	.473** .001 43	.711** .000 43

	R(仁)	Y(义)	L(礼)	Z(智)	X(信)	S1	S2	S3	S4	CA
Z Pearson Correlation	.518**	.482**	.676**	1	.695**	-.211	.068	.128	.695**	.555**
(智)Sig. (2-tailed)	.000	.001	.000		.000	.175	.665	.413	.000	.000
N	43	43	43	43	43	43	43	43	43	43
X Pearson Correlation	.357*	.317*	.473**	.695**	1	-.228	.051	.225	1	.564**
(信)Sig. (2-tailed)	.019	.038	.001	.000		.142	.745	.148		.000
N	43	43	43	43	43	43	43	43	43	43
S1 Pearson Correlation	-.052	-.145	-.255	-.211	-.228	1	.404**	-.678**	-.228	-.139
Sig. (2-tailed)	.741	.354	.098	.175	.142		.007	.000	.142	.372
N	43	43	43	43	43	43	43	43	43	43
S2 Pearson Correlation	.226	.139	.104	.068	.051	.404**	1	-.381*	.051	.230
Sig. (2-tailed)	.146	.376	.508	.665	.745	.007		.012	.745	.138
N	43	43	43	43	43	43	43	43	43	43
S3 Pearson Correlation	.153	.173	.248	.128	.225	-.678**	-.381*	1	.225	.150
Sig. (2-tailed)	.328	.267	.108	.413	.148	.000	.012		.148	.338
N	43	43	43	43	43	43	43	43	43	43
S4 Pearson Correlation	.051	-.096	.079	.077	.408**	-.142	.105	.154	.408**	.145
Sig. (2-tailed)	.744	.542	.616	.622	.007	.364	.502	.323	.007	.355
N	43	43	43	43	43	43	43	43	43	43
CA Pearson Correlation	.690**	.727**	.711**	.555**	.564**	-.139	.230	.150	.564**	1
Sig. (2-tailed)	.000	.000	.000	.000	.000	.372	.138	.338	.000	
N	43	43	43	43	43	43	43	43	43	43

** Correlation is significant at the 0.01 level (2-tailed).

* Correlation is significant at the 0.05 level (2-tailed).

1) *Testing hypothesis 1*

In the first hypothesis, it is hypothesized that in traditional Chinese community, there exist certain indicators of cognitive (community) social capital, which are commonly important in maintaining neighborhood social cohesion, despite the diverse social structures and groups of residents.To test the first hypothesis, the correlations among the five scales of cognitive social cohesion and that between cognitive social cohesion and structural social cohesion are the focuses.

- Hui Fang, Xi'an

As can be seen from the above results of the Pearson's correlation coefficient analysis(Table7-3), although representing social cohesion from different aspects,the nine scales are correlated with each other. With regards to the cognitive social cohesion, the five scales all have significant correlations with each other, which proves the validity of the cognitive social cohesion constructs—— Benevolence-Righteousness-Propriety-Wisdom-Faithfulness(仁-义-礼-智-信).

In terms of the structural social cohesion, scales are both negatively and positively correlated, and three scales indicating network integration, resource network, and participation of formal organizations have significant correlations with each other. Another scale representing the participation in informal groups is less correlated with the others. This result could be explained by the local residents' regular participation of formal organizations, which are mostly Mosques. Hence, the scales of network integration and resource network are highly correlated with local residents' participation of Mosques.

Regarding the correlations among cognitive social cohesion and structural social cohesion, only two scales of structural social cohesion, namely network integration and participation of formal organizations have significant correlations with the cognitive social cohesion scales, which to some extents, indicates that some structural social cohesion scales may not be highly influential to the entire cognitive social cohesion.

- Tianzi Fang, Shanghai

As can be seen from the above index (Table7-4), all the nine scales of social cohesion are correlated with each other. All the five scales of the cognitive social cohesion are highly correlated with each other, which also verified the validity of the cognitive social cohesion constructs. In terms of the structural social cohesion, scales are both negatively and positively correlated, and the interrelations of the four scales are not very clear in case of Tianzi Fang, which also indicate that structural social cohesion may not be very representative for ordinary local residents whose social relationship patterns remain as the traditional CHA-XU-GE-JU (差序格局).

Regarding the correlations among cognitive social cohesion and structural social cohesion, the above index indicates that the five scales of cognitive social cohesion are only highly correlated with one scale of structural social cohesion, namely participation of informal organizations, which to a large extent indicates that structural social cohesion scales may not largely influence the entire cognitive social cohesion. Moreover, the informal organizations could be mainly assumed as the "Tianzi Fang Shikumen Owner's Committee", as a majority of the residents agree to move out and rent their room for commercial uses.

In summary, firstly given that the validity of cognitive social cohesion constructs are verified in both cases, it could indicate that the cognitive social cohesion constructs are commonly influential in maintaining community social cohesion. Secondly, in both cases, there exist few significant correlations between cognitive social cohesion constructs and structural constructs, and only a few of the four scales of structural social cohesion have significant correlations with each other. Given that in both cases, the scales of cognitive social cohesion are highly correlated with each other, and that the interrelations within structural social cohesion vary according to cases and are not very clear, it indicates that the cognitive social cohesion constructs formed by cognitive social capital are mutually shared social values, and they are less influenced by the structural social cohesion constructs. As such, the first hypothesis could be successfully tested.

2) Testing hypothesis 2

To test the second hypothesis, which speculates that community social cohesion constructs have impacts on residents' perception on neighborhoods conservation, the focus is to examine the correlations of social cohesion constructs and residents' perceptions of neighborhood conservation in each neighborhoods.

- Hui Fang, Xi'an

As can be seen from Table 7-3, all the five scales of cognitive social cohesion are highly and positively correlated with the variable indicating local residents' attitudes towards community conservation. In addition, four out of five scales of cognitive social cohesion, namely Benevolence (仁),Righteousness(义),Propriety(礼) , and Wisdom (智), have significant correlations with local residents' attitudes towards community conservation. However, there are no significant correlations between the structural social cohesion and local residents' perspective towards community conservation. It can be summarized that there exists high correlation coefficient among the cognitive social cohesion and residents' perceptions of neighborhood conservation, with the following four scales——Benevolence, Righteousness, Propriety, and Wisdom being the most influential ones. In this way, it could partly tested hypothesis 2 in the case of Hui Fang in Xi'an.

- Tianzi Fang, Shanghai

As can be seen from Table 7-4, all five scales of cognitive social cohesion, namely Benevolence (仁),Righteousness(义),Propriety(礼) , Wisdom (智) ,and faithfulness(信) have significant correlations with local residents' attitudes towards community conservation. Regarding structural social cohesion, there are no significant correlations between the structural social cohesion and local residents' perspective towards community conservation.

As such, it can be summarized that residents' perceptions of neighborhood conservation are highly correlated with the cognitive social cohesion. In addition, the structural social cohesion constructs do not largely influence residents' perceptions of neighborhood conservation. In this way, hypothesis 2 could be partly tested in the case of Tianzi Fang in Shanghai.

In addition, a SWOT analysis could be applied in order to understand the strength and weakness of community social capital in both Hui Fang and Tianzi Fang, and also to analysis the opportunities and threats in front of the current development condition. For instance, in Hui Fang, regarding residents' attitude on conservation, both Justice(义) and Propriety(礼) have most significant correlations. While in Tianzi Fang, all the indicators have significant correlations. The strength of the social capital in Hui Fang lies in that it is highly closely bonded based on the social relationship developed mainly form Justice and Propriety, hence having the opportunities to bond closely and take collective actions in front of issues which will have significant influence to their entire ethnic community. Nevertheless, as it is not balanced due to the weakness of wisdom, Hui Fang will face the severe threats of unwise management or revitalization approach caused by lacking of wisdom and will not have many opportunities to break through the lock-in effect. On the other hand, as for Tianzi Fang, the strength of the community social capital lies in that it is more flexible in dealing with difficulties and hence more creative in terms of breaking lock-in effect.

7.2 Social network analysis in the decision making process

7.2.1 Hui Fang, Xi'an

1) *Identification and categorization of social relationship nodes in the network of decision making platform*

As shown in Table 7-5, during Xi'an Muslim Historic District Protection Project (1997-2002), 19 organizations/groups are identified as relationship nodes. The relationships ties among three specific nodes, namely Hua Jue Xiang Mosque, Residents of Renovated Courtyard Houses, and residents in general, are identified as bonding social capital ties within a community. Two nodes are identified as generating bridging social capital. Norwegian University of Science and Technology and Xi'an University of Architecture and Technology, as research institutes and professional consultancies, brought a horizontal network of resources to the community, as well as relationships from outside of the community. The Hua Jue Xiang Neighborhood Committee also functioned in connecting the community with different relationship resources, both official and non-official. The other 13 nodes are mostly governmental organizations and official associations, which could vertically link the community to authorities and powers and be identified as linking social capital. Among these nodes, the Xi'an Muslim Historic District Protection Project Office is active in keeping a two-way relationship with other government organizations, such as the Xi'an Municipal Construction Committee and the Xi'an Planning and Design Research Institute, through communication and collaboration.

Nodes producing social capital to Hui Fang during Xi'an
Muslim Historic District Protection Project (1997—2002)   Table 7-5

Social capital		Nodes of communication and cooperation producing social capital to community
Linking	1	Ministry of Science and Technology of China
	2	Department of Science and Technology Shaanxi Province
	3	Xi'an Municipal government
	4	Xi'an Municipal Construction Committee
	5	Xi'an Bureau of Housing and Land Administration 2nd Branch
	6	Xi'an Planning and Design Research Institute
	7	Xi'an Muslim Historic District Protection Project Office
	8	Norwegian Ministry of Foreign Affairs
	9	NORAD
	12	Beiyuanmen Street Office
	15	Xi'an Islamic Association
	16	Xi'an Ethnic Affairs Commission
	17	Construction Section
Bridging	10	Norwegian University of Science and Technology
	11	Xi'an University of Architecture and Technology
	13	Hua Jue Xiang Neighborhood Committee
Bonding	14	Hua Jue Xiang Mosque
	18	Residents of Renovated Courtyard Houses
	19	Neighborhood Residents

Referring to Table 7-6, 21 organizations/groups are identified as nodes of relationship ties during the Sajinqiao Redevelopment Project (2005—2007). The close relationship among three nodes, namely the Resident Representatives, Neighborhood Residents Volunteers, and Neighborhood Residents, are identified as bonding social capital since they are closely connected regarding the demolition issues within the Sajinqiao neighborhood. Five nodes, namely the West Mosque, the Old Mosque, Venerable Elders among the neighborhood, Residents in surrounding neighborhoods, and the 11 Mosques Union, are identified as bridging social capital by helping Sajinqiao Neighborhood connect with outside horizontal resources and relationships. The 11 Mosques Union and the West Mosque are especially active nodes in bridging the Sajinqiao Neighborhood with outside resources. Similar to Table I, the other 13 nodes are mostly governmental organizations and official associations that could provide the Sajinqiao Neighborhood with vertical links to authorities and powers, and may also be identified as generating linking social capital. In this case, the linking social capital found in the Sajinqiao Neighborhood ranges from the following four levels: the district level, the municipal level, the provincial level, and the central government level.

Nodes producing social capital to Hui Fang during Sajinqiao Redevelopment Project (2005—2007)   Table 7-6

Social capital		Nodes of communication and cooperation producing social capital to community
Linking	1	Central Government of China
	2	Shaanxi Provincial Government
	3	Shaanxi Provincial Ethnic Affairs Commission
	4	Xi'an Municipal Ethnic Affairs Commission
	5	Xi'an Municipal Islamic Association
	6	Xi'an City Planning Bureau
	7	Lianhu District Bureau of Construction
	8	Lianhu District Center of Development
	9	Lianhu District People's Procuratorate
	10	Lianhu District Public Security Bureau
	11	Lianhu District Bureau of Ethnic and Religious Affairs
	12	Miaohou Street Police Station
	13	Sajinqiao Neighborhood Committee
Bridging	14	The West Mosque
	15	The Old Mosque
	17	Venerable Elders among neighborhood
	20	Residents in surrounding neighborhoods
	21	The 11 Mosques Union
Bonding	16	Resident Representatives
	18	Neighborhood Residents Volunteers
	19	Neighborhood Residents

2) *Network analysis——Social network structure of decision making platform*
The social network structure of decision making platform during the conservation process of Hui Fang in Xi'an are illustrated in the following social network analysis of 1997-2002 (Figure. 7-1) and SRD 2005—2007 (Figure 7-2).

Data shows that the overall density of the entire relationship network of XMHDPP 1997—2002 (20.7%) is lower than in the entire network of SRP 2005-2007 (27.9%), and its ties of communication or cooperation (70) are fewer than that in SRP 2005-2007 (117)(Table 7-7).

These results indicate that, compared to SRP 2005—2007, there is a lack of intensive commu-

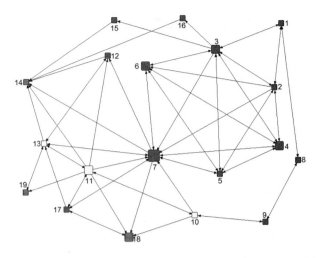

Figure 7-1 Network analysis of XMHDPP 1997—2002

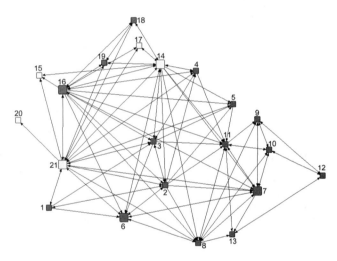

Figure 7-2 Network analysis of SRP 2005—2007

nication and cooperation among the interest groups in XMHDPP 1997—2002. Given that the density of communication and cooperation among linking social capital are considerably denser and some major linking social capital ranks top in degree centrality (12) (Table 7-8), it is assumed that a top-down pattern is dominant.

Compared to SRP 2005—2007, the weak degree of centrality of bonding social capital (degree=2) indicates that local residents might have not been very well informed and do not reveal a strong desire to participate. In SRP 2005—2007, a high ranking degree of centrality of both bonding (degree=14) and bridging social capital (degree=15) indicates a better participation pattern, which is highlighted as a positive function of social capital to promote collective actions.

Density overall of the conservation process in Hui Fang, Xi'an   Table 7-7

Projects	Density overall	
	Density	Number of ties
XMHDPP 1997—2002	0.2074	70
SRP 2005—2007	0.2786	117

Degree centrality of the conservation process in Hui Fang, Xi'an   Table 7-8

Nodes	XMHDPP 1997—2002	
	Degree	Nrm degree
Mean	4.737	26.316
Std Dev	2.53	14.059
Xi'an Muslim Historic District Protection Project Office	12	66
Xi'an University of Architecture and Technology	8	44
Neighborhood Residents	2	11
Nodes	SRP 2005—2007	
	Degree	Nrm degree
Mean	7.143	35.714
Std Dev	4.062	20.312
The 11 Mosques Union	15	75
The West Mosque	14	70
Resident Representatives	14	70

7.2.2　Tianzi Fang, Shanghai

1) *Identification and categorization of social relationship nodes in the network of decision making platform*

As Table 7-9 shows, 13 relationship nodes are identified in the initial stage. The relationships ties among Taikang Road Linong Factories are identified as bonding social capital, ties within the factory community. Six nodes are identified as generating bridging social capital. Zhang Jianjun as consultant and broker, Wu Meisen, Huang Yongyu, Chen Yifei and Er Dongqiang as important promoters, together with Taikang Road Artist Union as a formal organization of art business and general artists, brought to this place the horizontal network of resources and relationships from outside of the community. Zheng Rongfa functioned within the administrative system of Dapuqiao Sub-district, connecting Tianzi Fang Linong Factories with administrative power and resources. Luwan District Official Departments, Dapuqiao Sub-district Office(DSO), including Taikang Road Development Office(DSO), Administration Committee(DSO), Taikang Art Street Committee(DSO) functioned as official organizations, which vertically linked Taikang Road Linong Factories to specific authorities and powers and can be identified as generating linking social capital. Among these nodes, Dapuqiao Sub-district Office was active in keeping the two-way relationship ties with other official departments and social organizations. Due to the lock-in effect of administrative restructuring and economic development, Dapaqiao Sub-district Office functioned a very powerful role in promoting Taikang Road Art Project and Artist Factories.

Nodes producing social capital to Tianzi Fang in initial stage Table 7-9

Linking ■■■	Bridging ■■■	Bonding ■■■
1.Luwan District Official Departments	8.Mr. Zhang Jianjun	7.Linong factories
2. Mr. Zheng Rongfa	9.Mr. Wu Meisen	
3.Da Puqiao Sub-district Office,DPSO	10.Mr. Chen Yifei	
4.Taikang Road Development Office,DPSO	11.Mr. Er Dongqiang	
5.Administration Committee,DPSO	12. Mr. Huang Yongyu	
6.Taikang Art Street Committee,DPSO	13.Taikang Road Artist Union	

Referring to Table 7-10 in the second stage, total 21 organizations/groups and egos are identified as nodes of relationship ties. Ties of the close relationship among four nodes, namely, Zhou Xinliang as resident representative, residents who are in favor of the soft redevelopment approach, Residents who are opposed of the soft redevelopment approach, and Tianzi Fang Shikumen Owner's Committee as a bottom up resident organization are identified as bonding social capital, as they are closely connected with each other regarding the demolition issues and housing leasing issues within the residents community.

Eleven nodes, namely Zhang Jianjun, Wu Meisen, Li Wuwei and his Creative Research Project team , Ruan Yisan, Zheng Shiling and the Taikang Road Historic Street Conservation Project team of Tongji University, Taikang Road Management Committee organized by Wu Meisen as a informal organization to manage business and housing leasing issue, Tianzi Fang Intellectual Property Alliance, and Tianzi Fang association of Japanese company as nongovernmental organizations, and Shanghai Dingrong Real Estate Ltd,(ASE GROUP) as developer are identified as generating bridging social capital by connecting Tianzi Fang with outside horizontal resources and relationships. Especially Taikang Road Management Committee functions as active nodes bridging Tianzi Fang neighborhood with outside resources and opportunities.

Nodes producing social capital to Tianzi Fang in transitional stage Table 7-10

Linking	Bridging	Bonding
1. Mr. Zheng Rongfa	5. Shanghai Dingrong Real Estate Ltd.	10. Tianzi Fang Shiku Men Owner's Committee
2. Shanghai Government Departments	6. Mr. Zhang Jianjun	11.Tianzi Fang Residents (Opposed)
3.Luwan District Bureau of Planning	7. Mr. Wu Meisen	12.Tianzi Fang Residents (In favor of)
4. Dapuqiao Sub-district Office	8.Taikang Road managementCommittee	13.Mr. Zhou Xinliang
9. Jianzhong Community Committee	14.Taikang Road Historical Street Conservation team by Tongji University	
21. Luwan District Association of Science and Technology	15.Creative Research Project by Prof. Li Wuwei 16.Tianzi Fang Intellectual Property Alliance 17. Prof. Ruan Yisan 18.Prof. Zheng Shiling 19.Prof. Li Wuwei 20.Tianzi Fang Association of Japanese Enterprises	

The rest six nodes are mostly governmental organizations and official associations, which provided Tianzi Fang Neighborhood with vertical links to authorities and resources, and are identified as generating linking social capital. In this case, the linking social capital actually processed by Tianzi Fang Neighborhood range from the following three scales. the sub-district level and below includes Dapuqiao Sub-district office and Jianzhong Community Committee; the district level includes Luwan District Bureau of Planning, Luwan District Association of Science and Technology; and Shanghai municipal level include several Shanghai government departments.

Referring to Table 7-11, in the target stage, total 17 organizations and egos are identified. Similarly, ties of the close relationship among three nodes, namely, Zhou Xinliang as resident representative, residents who are in favor of Tianzi Fang, residents who are opposed of Tianzi Fang are identified as bonding social capital.

Eight nodes, namely Zhang Jianjun, Wu Meisen, Zhengrong Fa, Shanghai Huaxia Creative Research Institute, Tianzi Fang Trade Union, Research team led by Prof. Yu Hai, Research scholars and groups from Tongji University are identified as generating bridging social capital. Especially Tianzi Fang Trade Union and Shanghai Huaxia Creative Research Institute function as active nodes bridging Tianzi Fang neighborhood with outside resources and knowledge.

Nodes producing social capital to Tianzi Fang in target stage   Table 7-11

Linking	Bridging	Bonding
2. Shanghai Government Departments	1. Mr. Zheng Rongfa	11. Tianzi Fang Residents (Opposed)
3. Luwan District Bureau of Planning	5. Mr. Zhang Jianjun	12. Tianzi Fang Residents (In favor of)
4. Dapuqiao Sub-district Office	6. Mr. Wu Meisen	13. Mr. Zhou Xinliang
8. Jianzhong Community Committee	7. Tianzi Fang Trade Union	
9. Tianzi Fang Management Committee	14. Prof. Yuhai from Fudan University	
10. Tianzi Fang Property Management Office 16. Shanghai Tourism Administration	17. Shanghai Huaxia Creative Culture Research Institute	

The rest six nodes as official associations and organizations are identified as generating linking social capital, which provide Tianzi Fang with vertical links and resources to authorities and powers. Similarly, the linking social capital is Dapuqiao Sub-district office and Jianzhong Community Committee, Tianzi Fang management committee and Tianzi Fang Property management office, the Luwan District Bureau of Planning, and several Shanghai government departments such as Shanghai Tourism Administration.

2) *Network analysis——Social network structure of decision making platform*
The social network structure of decision making platform during the conservation process of Tianzi Fang in Shanghai are illustrated in the following stages of social network analysis (Figure.7-3).

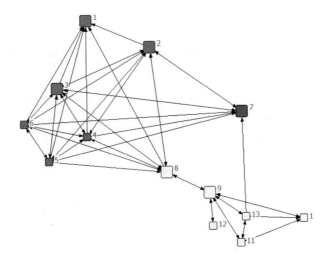

Figure 7-3   Initial stage of the conservation process in Tianzi Fang

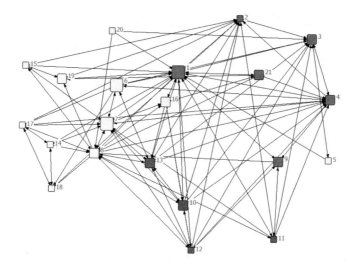

Figure 7-4 Transitional stage of the conservation process in Tianzi Fang

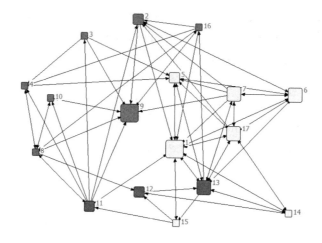

Figure 7-5 Target stage of the conservation process in Tianzi Fang

The network analysis of the initial stage indicates a top-down decision making pattern plus the horizontal participation of experts and stakeholders. It also illustrates a dense connection within linking social capitals, which explains why the overall density of its entire relationship network (40.38%) is higher than the other two stages (Table 7-12). Linking social capital is found to have played an important role in keeping the project implemented efficiently through top-down direction, nevertheless results also indicate that in the initial stage regarding decision making of Taikang Art street and artist factories, the decision was made collectively rather than a top-down choice. Seeking opportunities for development under the impact of Asian Economic Crisis, Dapuqiao Sub-district Office is well connected to other social resources and is open to good ideas from professional aspects. As can been seen in Table 7-9, aside from Dapuqiao Sub-district Office and Zhang Rongfa which could bring vertical links to authorities, Zhang Jianjun and Wu Meisen as egos to provided horizontal resources and knowledge are also very active in the decision making process, which brings new value to Linong Factories and leads to creative reuse.

Density overall of the conservation process in Tianzi Fang   Table 7-12

	Density	Number of Ties
Initial Stage	0.4038	63
Transitional Stage	0.3119	131
Target Stage	0.2941	80

The transitional stage represents a dynamic relationship network which is inclusive, accessible, open and balanced, as shown by degree centrality(Table 7-13). The highest mean degree centrality (8.19) and the actual number of ties (131) in the transitional stage indicate the intense communication and cooperation among the involved interest groups regarding the conservation of both factories and residential area against demolition. The lower overall density of the entire relationship network (31.19%) could be explained by the limitation to include considerable personal connections among influential people in political circles. Data indicates that Zheng Rongfa and Dapuqiao Sub-district Office, Wu Meisen and the Taikang Road Management Committee organized by Wu Meisen, Zhou Xinliang, the supportive residents, and their Tianzi Fang Shikumen Owner's Committee rank high in degree centrality, which means that linking social capital, bridging social capital and bonding social capitalare all active in the campaign to conserve Tianzi Fang based on different understandings of heritage value.

It is also assumed that good returns of soft redevelopment and conservation approach will affect local people, who will in turn work as strong social forces against demolition and trigger a breaking point of path dependency. In another word, the decision to conserve Tianzi Fang will meet the collective value of most interest groups and hence, promote social cohesion in the long run.

Degree centrality of the conservation process in Tianzi Fang   Table 7-13

Initial Stage	Degree	Number Degree
Mean	5.385	44.872
Top active nodes in initial stage		
Dapuqiao Sub-district Office	7.000	58.333
Mr. Zheng Rongfa	7.000	58.333
Mr. Zhang Jianjun	7.000	58.333

Initial Stage	Degree	Number Degree
Mr. Wu Meisen	5.000	41.667
Transitional Stage	Degree	Number Degree
Mean	8.190	40.952
Top active nodes in transitional stage		
Mr. Zheng Rongfa	18.000	90.000
Taikang Road Management Committee(By Wu)	16.000	80.000
Dapuqiao Sub-district Office	13.000	65.000
Mr. Wu Meisen	12.000	60.000
Mr. Zhou Xinliang	10.000	50.000
Tianzi Fang Residents (In favor of)	10.000	50.000
Tianzi Fang Shikumen Owner's Committee	9.000	45.000
Target Stage	Degree	Number Degree
Mean	6.118	38.235
Top active nodes in target stage		
Mr. Zheng Rongfa	10.000	62.500
Mr. Zhou Xinliang	9.000	56.250
Tianzi Fang Residents(Opposed)	8.000	50.000
Tianzi Fang Management Committee	7.000	43.750
Tianzi Fang Trade Union(by Wu)	7.000	43.750

Compared to previous stages, the lowest overall network density in the target stage together with the network pattern shows the target of building up the top-down decision making ever since Tianzi Fang has been taken over by government. High degree centrality of bonding social capital among residents who are against the soft redevelopment approach reveals that, these residents' interest and problems shall be considered in order to achieve social cohesion. A high ranking degree centrality of both bonding and bridging social capital indicates that even though there is a trend towards a top-down decision making pattern, which is dangerous when mass demolition is favored, again, with the active participation pattern of bonding and bridging social capital, the value of Tianzi Fang will be collectively decided and refined.

Implications highlight that the direct link/communication between nodes of linking and bonding as well as bridging social capital should be created to mitigate the strong top-down decision making process. Public intervention could be achieved through building up direct connections between community bonding and higher-level linking social capital. Feedbacks from professional, stakeholders and locals should have access to reshaping the value system of urban heritage before a top-down led demolition happens. Another implication addresses that in conservation, attention should be paid to build up intense connection and cooperation between bridging and community bonding social capital in order to generate better understanding and cooperation.

7.2.3 Testing hypothesis 3

Hypothesis 3 speculates that in the urban conservation decision-making network structure, there exist in some patterns of dynamic composition and interactions of different forms of social capital, which are influential to break the lock-in effect in decision making and to transform the conservation decision making towards a just decision-making process, thus creating new paths for urban conservation practices.

From the above social network analysis in both cases, it has revealed that: (1) during the initial stages, there exists a decision making network pattern dominant by linking social capital. This pattern is inclined to keep lock-in effect in decision making, which is to conduct mass demolition of historic neighborhoods and mass relocation of local residents and to gain land transaction profit; (2) during the transitional stages, there exist active participations, collaborations and interactions among local residents which generate bonding social capital and of several organizations which generate horizontal bridging social capital to the conservation process. In addition, data has shown that in the transitional stages, some active relationship nodes generating bonding and bridging social capital have both high overall density and degree centrality. These dynamic interaction and cooperation patterns of high overall degree and degree centrality of relationship nodes which generate bonding and bridging social capital within the decision making network structure, could be identified as the driving forces to break the top-down lock-in effect of conservation decision making, and to in turn transform the decision-making to conserve the community and to pursue collective profits; (3) In case of Tianzi Fang, there also exists linking social capital based governmental office which has very high density and degree centrality, and are well connected with other social capital form in transitional stage. It indicates that different relationship nodes generating three forms of social capital could all cooperate and interact actively if there are collective values among different interest groups; (4) Compared to the initial stage which is mainly top-down, in both transitional stages and target stages, there exist more inclusive patterns in terms of social capital form, especially the bonding and bridging social capital, which can be identified as a step forward to a just decision making network structure as well as a just decision making process. In light of the above discussions, the third hypothesis could be successfully tested.

Chapter 8 Conclusion

This closing chapter summarizes and concludes the research. It firstly starts with a brief review of this research, and moves on to summarize the main research findings. The contributions and implications of this doctoral study are discussed, which are followed by the discussions of research limitations as well as recommendations for future studies.

8.1 Research overview

Under the transitional social and economic contexts, recent decades have witnessed the evolving values of heritage, the shifting paradigms of urban conservation doctrines, and the emerging appeals for sustainable development, especially the dimension of social sustainability. With the intention to combine urban conservation, social sustainability and social capital, this study examines how social capital could contribute to urban conservation in order to achieve social sustainability.

What is the place of social capital in the process of urban conservation in historic quarters in order to achieve socially sustainable development? With the proposed research question, a gap of the social capital's role in urban conservation to achieve social sustainability is identified. Therefore, the main objective of this study is to reveal the role of social capital in urban conservation with regards to two key indicators of social values——social cohesion and social justice. Three hypotheses are proposed in order to examine the following relationships in two traditional Chinese neighborhoods: stable relationship based social capital and community social cohesion, social cohesion and residents' attitudes towards community conservation, as well as the dynamic composition and interactions of different forms of social capital in the network structures of conservation decision-making. The study employs a methodology which combines both qualitative and quantitative methods, and adopts case study, social capital survey and social network analysis under the framework of path dependency as main approaches to facilitate this research.

8.2 Discussions of main findings

8.2.1 Neighborhood social cohesion constructs in Chinese cities

Based on the social capital survey in two Chinese traditional neighborhoods, the five scales of the cognitive social cohesion constructs, namely five cognitive social capital "Benevolence"(仁),

"Righteousness"（义）, "Propriety"（礼）, "Wisdom"（智）, and "Faithfulness"（信）are generally highly correlated with each other. Structural social cohesion takes different influential forms in different neighborhoods. In some neighborhood there are close networks of informal organizations, while in some neighborhood it is the networks of formal organizations that are influential. Nevertheless in different neighborhoods with different social structures, the five scales of cognitive social cohesion remain highly correlated with each other, which indicates that in traditional Chinese neighborhoods, the cognitive social cohesion constructs formed by stable social relationships (cognitive social capital) are commonly influential in maintaining community social cohesion. Given that all the five cognitive social cohesion constructs are developed based on a perspective of "CHA-XU-GE-JU"（差序格局）, it also suggests that in traditional Chinese neighborhoods, the traditional relationship pattern still plays influential role in shaping people's social relationships and their commonly shared social values——social cohesion（和）in their daily lives.

8.2.2 Neighborhood social cohesion and residents' attitudes of community conservation

In both cases, residents' perceptions of neighborhood conservation are mostly highly correlated with the cognitive social cohesion. In addition, the structural social cohesion constructs do not largely influence residents' perceptions of neighborhood conservation. It indicates that cognitive community social cohesion plays a significant role in strengthening resident's attitudes to conserve their community. Furthermore, in case of Chinese traditional neighborhoods, as the cognitive social cohesion constructs encompass the social values such as mutual help and supports, to a certain extent living in traditional community with a sense of social cohesion "would mitigate social disparity, injustice, and tension"(Cheung & Leung, 2011, p. 565), thus reducing the social cost and the burdens to the city. In addition, this finding also reflects that cognitive community social cohesion could be understood as mutual social values attached to traditional neighborhood, and should be added into the consideration of urban conservation planning decision making, especially in Chinese cities.

8.2.3 The dynamic compositions and interactions initiated by three forms of social capital in the decision-making network

Based on the case studies in two Chinese traditional neighborhoods, the functions of three forms of social capital, namely bonding, linking, and bridging social capital in the conservation process are identified to be similar with what has been widely discussed (Bank, 2000; Putnam, 2000). What drew the attention lies in the dynamic compositions and interactions of three forms of social capital in the inclusive decision-making network, which have been highlighted in the discussions of social justice (Healey, 2003; Young, 1990).

Based on social network analysis under the path dependency framework, it has revealed in both cases that: given other driving forces remaining the same, (1) in the initial stage, if a decision making network pattern isdominated by single social capital form, this pattern is inclined to keep lock-in effect in decision making[①]; (2) during the transitional stages, if there exist active participations, collaborations and interactions which generate different forms of social capital, the lock-in effect are likely to be broken and new path might be created, and these compositions of social capital would be understood as the driving forces (Healey, 2003) to break the lock-in effect of conservation decision making, and to negotiate the decision making towards the target stage. Moreover, the dynamic composition and interaction of different social capital in the decision making network could be measured by social network parameters: overall degree and degree centrality of relationship nodes; (3) Different relationship nodes generating three forms of social capital could all cooperate and interact actively if there are collective values among different interest groups, as is revealed in case of Tianzi Fang; (4) Compared to the initial stage, in both transitional stages and target stages, if there exist more inclusive patterns in terms of social capital form and diverse interest groups, especially the bonding and bridging social capital, it could be identified as a step forward to a just decision making network structure as well as a just decision making process.

Regarding this, it must be highlighted that in case of Hui Fang in Xi'an, though the West Mosque and the Mosque Union are categorized as bridging social capital, the exterior resources they could bring to the community are fairly limited to a certain kinds. In this case, though they are quite active in terms of network density and degree centrality, after the transitional stage, the community suffers from a further decay. On one hand, there are no efforts from local government, on the other hand, there lack efforts initiated by "real" bridging social capital which could function well to bring creative ideas and many kinds of resources to revitalize this closely bonded community. In light of this, the functions of different forms of social capital must be thoroughly evaluated, aside from the network structure analysis.

8.3 A social capital integrated approach to urban conservation

8.3.1 The place of social capital in the process of urban conservation of historic quarters in order to achieve social sustainability

This discussion is elaborated based on the two indicators of social sustainability: social cohesion and social justice. In the first place, this study has proved that social capital, especially the

---

① In the two research cases, decision-making is dominant by linking social capital and the lock-in effect means to conduct top- down mass demolition of historic neighborhoods and mass relocation of local residents, with a returns of land transaction profit.

cognitive social capital, which are formed by long-term stable relationships within traditional neighborhoods are commonly influential in maintaining community social cohesion.

In the second place, highly correlated with local residents' attitudes of community conservation, cognitive community social cohesion constructs could be understood as mutual social values attached to traditional neighborhood. Therefore, cognitive social capital formed by long-term stable relationships plays significant roles in shaping the mutual social values attached to traditional neighborhoods. In addition, as mentioned above, cognitive social cohesion as mutual social values attached to traditional neighborhood, which enhance the affordable living and sense of belonging should be added into the consideration of urban conservation planning decision making. In light of this, it could contribute to social justice in terms of what Healey (2003, p. 105)called "moral commitment".

In the third place, as has been discussed, in the conservation decision-making network structure, the dynamic composition and interactions initiated by different forms of social capital will contribute to social justice in terms of an inclusive decision-making network structure.

In the fourth place, as has been proved in two cases, from the initial stage which is normally dominated by lock-in effects, social capital initiated participations and interactions could negotiate the urban conservation decision making, and as driving forces of breaking point, could break lock-in effect and transit the decision making towards new development path. In light of this, social capital could contribute to social justice in terms of conservation decision making process.

8.3.2 A social capital integrated approach to urban conservation

The above discussions reveal the roles of social capital in urban conservation in order to achieve social sustainability. As social capital has been proved to be significantly related with the two indicators of social sustainability ——social cohesion and social justice, it is necessary to adopt social capital into the field of urban conservation. This study proposes a social capital integrated approach to urban conservation.

This approach combines several methods, including social capital survey, social network analysis, and path dependency analysis, etc. By applying this approach the field of urban conservation could be advanced in terms of understanding core social values attached to traditional neighborhoods, enhancing community social cohesion and strengthening residents' awareness to community conservation, optimizing decision-making network structure, and negotiating decision-making towards a just process, and so on and so forth.

In addition, regarding urban heritage conservation, which is understood as the value-centered place management process, this social capital integrated approach could facilitate urban heritage conservation in the following aspects: 1) understanding the mutual social values attached to heritage; 2) providing an inclusive decision making platform for heritage values to be collectively defined from different interest groups and stakeholders, and this decision making platform is in line with what Pendlebury (2008) has defined as "opportunity spaces" ; 3)negotiating the heritage conservation process towards a just process of place management; 4) bringing more creative ideas into heritage conservation , so that compared to the "no choice" situation which is usually dominated by lock-in effect, people could see the alternatives and make a choice .

8.4 Contributions and implications

8.4.1 Theoretic contributions

The findings of this study could extend the body of knowledge on urban conservation in terms of social sustainability through the adoption of a comprehensivesocial capital approach.On one hand,through social capital survey this study indicates that social capital plays an important role in constructingthe community social cohesion.On the other hand, through the social network analysis certain patterns of compositions and dynamic interactions initiated by different forms of social capital have been identified, which could contribute to transform or create the path towards inclusive decision-making pattern, thus contributing to social justice in a process perspective.

8.4.2 Methodological contributions

As has been mentioned in the discussion of social capital approach, this social capital integrated approach could facilitate urban heritage conservation in the following aspects: 1) value identification;2) building up an inclusive decision making platform;3)negotiating the heritage conservation process; and 4) breaking lock-in effect.

8.4.3 Practical implications

Based on the findings, this research proposes a social capital integrated approach to urban conservation, which has the following practical implications. Firstly, this approach could facilitate in providing sustainable urban conservation strategies in terms of heritage value selection and decision making process. Especially as the study explores the traditional residential neighborhoods in Chinese cities, the analysis of social value and decision-making process could be applied to similar traditional communities in China; secondly, for on-going urban conservation planning, this research could provide evaluation method for planning boards to review and evaluate the impacts of urban conservation outcomes and decision making process, thus contributing in modifying and adjusting existing urban conservation policies and transiting the path of urban conservation decision-making towards sustainable direction; thirdly, for potential urban conservation plan, this approach would facilitate the planning authorities to identify the community social values and to establish an inclusive decision-making platform, to build up the capacity of social capital raising, and to promote efficient public participation, hence improve urban conservation planning practices in terms of value selection, decision making process and planning outcome in the long run.

Moreover, social capital includes social relationship, social values, network and resources, etc. In future practices, a SWOT analysis of social capital could be adopted in order to provide insights on how to enhance the strength and opportunity of certain social capital, and to overcome weakness and avoid the threat wisely.

8.5 Limitations of the study

This piece of research work has its limitations. The first limitation lies in the limited number of cases. As this research employs an inductive method to explore the role of social capital on a case by case base, a large number of cases could add more validity to the results. However, due to the availability of cases and data collection, this research only focuses on two cases. In future studies, more case studies will be conducted.

Besides, in order to grasp the mutual qualitative character of social capital among these two different cases, this research did not compare the diverse urban contexts and analyze the difference of each community such as religion, culture and social backgrounds. Especially regarding the cognitive social cohesion constructs, which is assumed to be commonly shared as a core social values within community, there lacks a consideration of the impacts of religion, cultural diversity and other qualitative influential factors, which will have different influence accordingly. Given that this research aims to firstly propose a common social cohesion constructs, this research leaves all the other influential factors out of account. In future studies, a more subtle analysis considering these influential factors will be conducted.

Furthermore, compared to cognitive social capital, the structural aspects of social capital are not the focus of this study, hence the structural aspects of social cohesion has not been fully examined. In future study, this limitation should be considered.

With regards to the cognitive social cohesion constructs, which is developed from the traditional relationship pattern –CHA-XU-GU-JU (差序格局), there lacks the consideration of the influence from modern social relationship pattern. The argument lies in that tradition is considered to be stable and not easy to be changed. However, in future studies, the influence of modern relationship pattern should be considered.

In addition, regarding to the social network analysis, this research only adopts overall degree and degree centrality as parameters. In future study, a more complicated group of parameter should be applied in order to gain a more comprehensive understanding of network patterns.

8.6 Conclusion

Nowadays, with stronger motivations to enhance the economic development and improve the physical environment of urban areas, social sustainability has remained ignored. In the advent of a worldwide consensus that urban conservation are getting increasingly important, research on urban conservation should be encouraged to adopt the sustainable development perspective. Especially the focuses on the social dimension will be significant topics that merit attention by researchers engaged in urban conservation studies.

Although this research has many limitations and is far from providing practical models or solutions for urban conservation, it takes the initiatives to discuss the role of social capital in urban conservation in order to promote socially sustainable development. This work could be considered as the preliminary study to reveal the role of social capital in urban conservation and to develop a social capital integrated approach for socially sustainable urban conservation.

References

Adler, P. S. and S. W. Kwon (2002). "Social capital: Prospects for a new concept." Academy of management review: 17-40.

Ahuja, G. (2000). "Collaboration networks, structural holes, and innovation: A longitudinal study." Administrative science quarterly 45(3): 425-455.

Akagawa, N. (2014). Heritage Conservation and Japan's Cultural Diplomacy: Heritage, National Identity and National Interest, Routledge.

Alexander, A. (2009). Britain's new towns: garden cities to sustainable communities, Taylor & Francis.

Alguezaui, S. and R. Filieri (2010). "Investigating the role of social capital in innovation: sparse versus dense network." Journal of knowledge management 14(6): 891-909.

Al-hagla, K. S. (2010). "Sustainable urban development in historical areas using the tourist trail approach: A case study of the Cultural Heritage and Urban Development (CHUD) project in Saida, Lebanon." Cities 27(4): 234-248.

Appleyard, D. (1979). The conservation of European cities, MIT Press Cambridge.

Ardakani, M. K. and S. S. A. Oloonabadi (2011). "Collective memory as an efficient agent in sustainable urban conservation." Procedia Engineering 21: 985-988.

Ashworth, G. (1984). "The management of change: Conservation policy in Groningen, The Netherlands." Cities 1(6): 605-616.

Ashworth, G. and J. Tunbridge (1990). The tourist-historic city, Belhaven Press, London.

Ashworth, G. and J. Tunbridge (2000). The tourist-historic city: retrospect and prospect of managing the heritage city (advances in tourism research series), Oxford, UK: Pergamon.

Ashworth, G. J. (1991). Heritage planning: conservation as the management of urban change, Geo Pers The Netherlands.

Atkinson, R. (2000). "The hidden costs of gentrification: displacement in central London." Journal of housing and the built environment 15(4): 307-326.

Atlee, T. (2008). " Principles of Public Participation."

Avrami, E., R. Mason, et al. (2000). "Values and heritage conservation." Conservation: the Getty Conservation Institute newsletter 15(2): 18-21.

Babaei, H., N. Ahmad, et al. (2012). "Bonding, Bridging, and Linking Social Capital and Psychological Empowerment among Squatter Settlements in Tehran, Iran." Journal of Basic and Applied Scientific Research 2(3): 2639-2545.

Baker, W. E. (2000). Achieving Success through Social Capital: Tapping the Hidden Resources in Your Personal and Business Networks. University of Michigan Business School Management Series, San Francisco: Jossey-Bass.

Bank, W. (2000). World Development Report, 2000/2001: Attacking Poverty, Oxford University Press New York.

Bao, L., S. Qin, et al. (2005). "A Comparative Study on Residents, Perceptions of Tourism Impactson Traditional Ancient Dwell Houses Tourism Area——A case study of Xidi and Zhouzhuang [J]." Journal of Wanxi University 2.

Barrett, H. (1993). "Investigating townscape change and management in urban conservation areas: the importance of detailed monitoring of planned alterations." Town Planning Review 64(4): 435.

Berge, E. (2007). Social Capital: basic theory and role in economic development

Berger-Schmitt, R. (2000). Social cohesion as an aspect of the quality of societies: concept and measurement, ZUMA.

Bhatta, K. D. (2009). "Urban Heritage Conservation and Sustainable Community Development: A Case Study of Historic Town Thimi, Nepal." Construction of Matsapha International Airport in Swaziland, Southern Africa and Its Effect on Environmental and Social Ecology-Sk Dev 3: 7.

Bi, J. (2004). Reuse of the Courtyard Homes Protected as the Heritage in Drum Tower District of Xi"an, Xian University of Architecture and Technology.

Bian, Y. (1997). "Bring Strong Ties Back in: Indirect Ties, Networks Bridges, and Job Searches in China." American sociological review 62: 266-285.

Bian, Y. and S. Ang (1997). "Guanxi networks and job mobility in China and Singapore." Social Forces 75(3): 981-1005.

Blokland, T., T. Blokland, et al. (2008). Networked urbanism: social capital in the city, Ashgate Pub Co.

Blundell, D. (1992). Tourism resources and cultural preservation: Comparison on traditional contemporary and prehistoric heritage in Taiwan. International Symposium on Tourism and Leisure Management. Taipei: Pacific Cultural Foundation, the Tourism Society, and the Graduate School of Tourism, Chinese Culture University. May 21st-24th.

Borgatti, S. P., M. G. Everett, et al. (1999). "UCINET 5.0 Version 1.00." Natick: Analytic Technologies.

Bourdieu, P. (1986). The forms of capital, Wiley Online Library.

Bourdieu, P. and L. J. D. Wacquant (1992). An invitation to reflexive sociology, University of Chicago Press.

Brimblecombe, P. and C. Saiz-Jimenez (2004). "Damage to cultural heritage." Air Pollution and Cultural Heritage: 87-90.

Buchanan, C. (1968). Bath: A Study in Conservation, Her Majesty's Stationery Office.

Burt, R. S. (1995). Structural holes: The social structure of competition, Harvard Univ Pr.

Burt, R. S. (2004). "Structural holes and good ideas." American journal of sociology 110(2): 349-399.

Burtenshaw, D., M. Bateman, et al. (1991). The European city: a Western perspective, Fulton.

Cao, W. (2005). Study on the renewal building in the historic drum-tower district of Xi'an [西安鼓楼历史街区更新建筑研究], Xi'an University of Architecture and Technology.

Cardoso, R. and I. Breda–Vazquez (2007). "Social justice as a guide to planning theory and practice: analyzing the Portuguese planning system." International Journal of Urban and Regional Research 31(2): 384-400.

Castells, M. (1985). The city and the grassroots: a cross-cultural theory of urban social movements, Univ of California Pr.

Castells, M. (1985). The city and the grassroots: a cross-cultural theory of urban social movements, Univ of California Pr.

Castells, M. and P. G. Hall (1994). Technopoles of the world: the making of twenty-first-century industrial complexes, Routledge.

Chang, Q. (2003). A Conservation Strategy of Architectural Heritage, 同济大学出版社.

Chang, T. (2000). "Singapore's Little India: A tourist attraction as a contested landscape." Urban Studies 37(2): 343.

Chang, T. and B. S. Yeoh (1999). "'New Asia–Singapore': communicating local cultures through global tourism." Geoforum 30(2): 101-115.

Charter, A. (1931). The Athens Charter for the Restoration of Historic Monuments. the 1st Int. Congress of Architects and Technicians of Historic Monuments, Athens, Greece.

Charter, V. (1964). International charter for the conservation and restoration of monuments and sites. IInd International Congress of Architects and Technicians of Historic Monuments, Venice.

Chase, L. D. (2000). "Saving place, municipal government and heritage conservation: the case of the Mount Newton Valley District of Central Saanich, British Columbia."

Chen, F. (2010). "Traditional architectural forms in market oriented Chinese cities: Place for localities or symbol of culture?" Habitat international.

Chen, F. (2011). "Traditional architectural forms in market oriented Chinese cities: Place for localities or symbol of culture?" Habitat international 35(2): 410-418.

Chen, S. (2005). Study on Conservation of the Dwelling culture and Living Environment of Xi'an Muslim Concentrated Area, Xian University of Architecture and Technology.

Cheng, M.-T. and C.-C. Fu (2010). From Breweries to Creative & Cultural Parks–The Challenge and Potentialities of Industry Heritage in Taiwan. the Proceeding of "Reusing the Past– ICOHTEC, TICCIH & Worklab Joint Conference.

Cheung, C.-k. and K.-k. Leung (2011). "Neighborhood homogeneity and cohesion in sustainable community development." Habitat International 35(4): 564-572.

Chiu, C. and C. F. Moss (2007). "The role of the external ear in vertical sound localization in the free flying bat, *Eptesicus fuscus*." Journal of the Acoustical Society of America 121(4).

Christiaan, G., D. Narayan, et al. (2004). "Measuring Social Capital. An Integrated Questionnaire." World Bank Working Paper(18).

Cohen, D. and L. Prusak (2001). In good company: How social capital makes organizations work, Harvard Business Press, Boston.

Cohen, N. (1999). Urban conservation. Cambridge, Mass., MIT Press.

Coleman, J. S. (1988). "Social capital in the creation of human capital." American journal of sociology: 95-120.

Coleman, J. S. (1994). Foundations of social theory, Belknap Press, Cambridge, MA.

Collins, R. C. (1980). "Changing views on historical conservation in cities." The ANNALS of the American Academy of Political and Social Science 451(1): 86-97.

Congress, U. (1966). "National Historic Preservation Act 1966."

Conzen, M. (1973). Geography and townscape conservation.

Conzen, M. R. G. (1960). "Alnwick, Northumberland: a study in town-plan analysis." Transactions and Papers (Institute of British Geographers)(27).

Côté, S. and T. Healy (2001). The well-being of nations: The role of human and social capital, Centre for Educational Research and Innovation.

Couch, C. and A. Dennemann (2000). "Urban regeneration and sustainable development in Britain:: The example of the Liverpool Ropewalks Partnership." Cities 17(2): 137-147.

Creighton, J. L. (2005). The public participation handbook: Making better decisions through citizen involvement, Jossey-Bass Inc Pub, San Francisco.

Cullen, G. (1961). The concise townscape, London: Architectural Press.

Cullingworth, J. B. (1999). British planning: 50 years of urban and regional policy, Bloomsbury Publishing.

Cuthill, M. (2010). "Strengthening the 'social' in sustainable development: Developing a conceptual framework for social sustainability in a rapid urban growth region in Australia." Sustainable Development 18(6): 362-373.

De Roest, H. and H. Noordegraaf (2009). " We learned it at our mothers' knees". Perspectives of churchgoing volunteers on their voluntary service'." Reformed World 59(3): 213-226.

Deng, Q. (2006). ""The origin, development, and its basic connotation of "Benevolence-Righteousness-Propriety-Wisdom-Faithfulness" (I) [" 仁义礼智信 " 的由来，发展及其基本内涵（上）]." Journal of Changsha University 19(6): 4-10.

Deng, Q. (2006). "The origin, development, and its basic connotation of "Benevolence-Righteousness-Propriety-Wisdom-Faithfulness" (II) [" 仁义礼智信 " 的由来，发展及其基本内涵（下）]." Journal of Changsha University 20(1): 1-5.

Ding, H. (2005). "About the Issue of Islam's Acclimatizing Itself to China's Society: From the Angle of Cultural Identification of Hui [J]." NorthWest Minorities Research 2: 006.

Dong, J. (2004). Urban histoy of China Beijing, China Architecture & Building Press.

Dong, W. (1996). "Free market economy driven by urban change--self-built renewal in Hui communites, Xi'an [自由市场经济驱动下的城市变革——西安回民区自建更新研究初探]." Urban Planning: 42-45.

Doratli, N. (2005). "Revitalizing historic urban quarters: A model for determining the most relevant strategic approach." European planning studies 13(5): 749-772.

Doratli, N., S. O. Hoskara, et al. (2004). "An analytical methodology for revitalization strategies in historic urban quarters: a case study of the Walled City of Nicosia, North Cyprus." Cities 21(4): 329-348.

Editing-Group (2009). Brief History of Hui Ethnic Beijing, The Ethnic Publishing House.

Elsorady, D. A. (2011). "Heritage conservation in Rosetta (Rashid): A tool for community improvement and development." Cities.

Environment, B. C. R. T. o. t. (1994). Public Involvement in Government Decision-making: Choosing the Right Model: a Report of the BC Round Table on the Environment and the Economy, The Round Table.

Evans, J. and P. Jones (2008). "Rethinking sustainable urban regeneration: ambiguity, creativity, and the shared territory." Environment and Planning A 40(6): 1416.

Fainstein, S. (1997). "Justice, politics and the creation of urban space." The urbanization of injustice: 18-44.

Fainstein, S. (2005). "Planning theory and the city." Journal of Planning Education and Research 25(2): 121-130.

Fan, W. (2004). the conservation and renewal of lilong housing in Shanghai [上海里弄的保护与更新], Shanghai Scientific and Technical Publishers.

Fang, G. (1996). "唐代外事机构论考." Journal of Chinese Historical Studies 1996 2: 8.

Fei, X. (1947). Xiang Tu Zhong Guo, 上海观察社.

Fei, X. (1989). "the Pattern of "Unity in Diversity" of China[中华民族的多元一体格局]."

Journal of Peking University[Humanities and Social Sciences 4(1.19)].

Fei, X. (1992). From the soil, the foundations of Chinese society: a translation of Fei Xiaotong's Xiangtu Zhongguo, with an introduction and epilogue, Univ of California Press.

Feilden, B. M. and J. Jokilehto (1993). Management guidelines for world cultural heritage sites, ICCROM.

Field, J. (2003). Social capital. London, Routledge.

Field, J. (2003). Social capital: Key ideas, Routledge.

Field, J. (2008). Social capital, Routledge.

Franke, S. (2005). Measurement of social capital: Reference document for public policy research, development, and evaluation, Policy Research Initiative.

Friedkin, N. E. (2004). "Social cohesion." Annual Review of Sociology: 409-425.

Fu, S. (1996). "The historical urban area protection (international) conference held in Huangshan city." ARCHITECTURAL JOURNAL.

Fu, X. (2001). The ancient Chinese city planning, the architectural complex layout and architectural design method research [中国古代城市规划,建筑群布局及建筑设计方法研究], China architecture & building press.

Fukuyama, F. (1996). Trust: The social virtues and the creation of prosperity, Free Pr.

Fukuyama, F. and I. Institute (2000). Social capital and civil society, International Monetary Fund, IMF Institute Washington, DC.

Galston, W. A. (2001). "Bowling Alone: The Collapse and Revival of American Community." Journal of Policy Analysis and Management 20(4): 788-790.

Galway, N. and M. Mceldowney (2006). "Place and Special Places: Innovations in Conservation Practice in Northern Ireland." Planning Theory & Practice 7(4): 397-420.

Ge, C. (1998). The Culture of Qin-Long [秦陇文化志]. Shanghai, Shanghai People's Publishing House.

Geddes, P., R. T. LeGates, et al. (1949). Cities in evolution, Williams and Norgate London.

Gehl, J. (2011). "Life between buildings."

Ghosh, S., R. Gupta, et al. (1996). Architectural and urban conservation, Centre for Built Environment.

Gibson, C. and L. Kong (2005). "Cultural economy: a critical review." Progress in human geography 29(5): 541-561.

Gittell, R. J. and A. Vidal (1998). Community organizing: Building social capital as a development strategy, Sage Publications, Inc.

Glaeser, E. L., D. Laibson, et al. (2002). "An Economic Approach to Social Capital*." The Economic Journal 112(483): F437-F458.

Gottdiener, M. and L. Budd (2005). Key concepts in urban studies, Sage Publications Ltd, London.

Gottdiener, M. and R. Hutchison (1994). The new urban sociology, McGraw-Hill New York.

Gratz, R. B. (1995). The living city: How America's cities are being revitalized by thinking small in a big way, Wiley.

Grootaert, C. (2004). Measuring social capital: an integrated questionnaire, World Bank Publications.

Gu, C. and G. Song (2001). "THE STUDY ON THE URBAN IMAGE AND ITS APPLICATION IN THE URBAN PLANNING [J]." City Planning Review 3.

Gu , K. (2001). "Urban morphology of China in the post-socialist age: Towards a framework for analysis." Urban Design International, 6 3(4): 125-142.

Guan, J. (2008). The Operation Mechanism Development of Urban Regeneration in Shanghai

Downtown, Tongji University.

Guo, H. (1983). "Issues of absolutism centralized highly fortified in Ming dynasty [关于明代专制主义中央集权高度强化的问题]." Journal of Northwest Normal University (Social sciences).

Guo, X. (2012). "Heritage conservation in urban renew [城市更新中的遗产保护]." Journal of North China Inistitute of Aerospace Engneering: 3-5.

Guo-zhao, L. (2007). "A Study on Taipingqiao Area and Xintiandi Based on Preservation Legislations." Time+ Architecture.

Guthrie, D. (1998). "The declining significance of guanxi in China's economic transition." The China Quarterly 154: 254-282.

Güzey, Ö. (2009). "Urban regeneration and increased competitive power: Ankara in an era of globalization." Cities 26(1): 27-37.

Hai, Y. (2008). "Narration of Historic Block Renovation in Power and Concept Dimensions: Case of Tianzifang in Shanghai."

Hai, Y. (2009). "Tianzifang Experiment: the City Renewal Model Superseding the Binary Opposition of a Place." China Ancient City.

Halpern, D. (1999). "Social capital: the new golden goose." London: Institute for Public Policy Research.

Han, J., L. Wang, et al. (2015). Addressing Social Revitalization in Conservation of Historic Quarters in China. Building Resilient Cities in China: The Nexus between Planning and Science, Springer: 75-91.

Han, Y. (2003). "The Silk Road and the introduction of Islam in northwest of China in Tang Dynasty [" 丝绸之路 " 与唐代伊斯兰教传入西北]." Journal Of Qinghai Nationalities Institute (social science): 59-63.

Hancock, T. (2009). Social Sustainability: The "soft infrastructure"of a healthy community, Retrieved March.

Hao, M. (2009). Studies on certain questions in the history of Hui in Xi'an [西安回族的若干历史问题研究], Shaanxi Normal University.

Harvey, D. (1973). Social justice and the city, Hodder Arnold, London.

Harvey, D. (1989). The urban experience, Johns Hopkins University Press Baltimore, MD.

Harvey, D. (1992). "Social Justice, Postmodernism and the City*." International Journal of Urban and Regional Research 16(4): 588-601.

Harvey, D. (2003). "The right to the city." International Journal of Urban and Regional Research 27(4): 939-941.

Haughton, G. and C. Hunter (1994). Sustainable cities, J. Kingsley Publishers.

He, J. (1980). "The evolution of Shi-Fang planning system in Tang and Song Dynasties [唐宋市坊规划制度演变探讨]." ARCHITECTURAL JOURNAL.

He, S. and F. Wu (2005). "Property-led redevelopment in post-reform China: a case study of Xintiandi redevelopment project in Shanghai." Journal of Urban Affairs 27(1): 1-23.

Healey, P. (2003). "Collaborative planning in perspective." Planning theory 2(2): 101-123.

Healey, P. (2006). "Transforming governance: challenges of institutional adaptation and a new politics of space 1." European planning studies 14(3): 299-320.

Henderson, J. (2000). "Attracting tourists to Singapore's Chinatown: a case study in conservation and promotion." Tourism Management 21(5): 525-534.

Henderson, J. (2003). "Ethnic heritage as a tourist attraction: The Peranakans of Singapore." International Journal of Heritage Studies 9(1): 27-44.

Heng, C. (1995). Urban conservation in fast growing historic cities. Planning in a Fast Growing Economy, Proceedings of the Third International Congress of the Asian Planning Schools Association, Sinagpore.

Heng, C. K. (1999). Cities of aristocrats and bureaucrats: the development of medieval Chinese cityscapes, Singapore Univ Pr.

Heng, C. K. and V. Chan (2000). "The" Night Zone" Storyline: Boat Quay, Clarke Quay and Robertson Quay." Traditional Dwellings and Settlements Review: 41-49.

Heng, C. K. and B. L. Low (2009). "New Asian public space: Layered Singapore." Urban Design International 14(4): 231-246.

Heng, C. K. and C. E. Quah (2000). "REINVENTING SINGAPORE'S CHINATOWN." Traditional Dwellings and Settlements Review: 58-58.

Henrichsen, C. (1998). "Historical outline of conservation legislation in Japan." Hozon: Architectural and Urban Conservation in Japan: 12-21.

Hisa, C.-J. (1998). "Historic Preservation in Taiwan:A Critical Review/ 台湾的古迹保存：一个批判性回顾." Journal of Building and Planning/ 国立台湾大学建筑与城乡研究学报 9: 1-9.

Hobson, E. (2004). Conservation and planning: changing values in policy and practice, Spon Press, London/New York.

Hoyem, H. (2004). "Path Dependency-A Useful Method for Planning and Design of Historical District ? ." Community(2): 75-81.

Hu, L. (2005). "Aroun M50 in Moganshan Road [在莫干山路 50 号流连]." Shanghai Wave(10): 41-45.

Hu, R. (2008). "Luwan District Local History 1994-2003."

Huang, J. (2008). "Exploration and Practice of the Institional Innovations of Shanghai Governnance Reform in 1990s." Shanghai Party History and Construction 6: 004.

Huang, J. (2010). Study on the Construction and Morphology Transition of the Drum-tower Hui Community in Xi'an [西安鼓楼回族聚居区结构形态变迁研究], South China University of Technology.

Huang, Q. (2008). The Protection and Reuse of Modern Industrial Buildings in Shanghai [上海

近代工业建筑保护和再利用], Tongji University.

Huisman, M. and M. A. J. Van Duijn (2005). "Software for social network analysis." Models and methods in social network analysis: 270-316.

Hung-Jen, T. and P. Waley (2006). "Planning through procrastination: The preservation of Taipei's cultural heritage." Town Planning Review 77(5): 531-555.

Huntington, S. P. (1971). "The change to change: modernization, development, and politics." Comparative Politics: 283-322.

HÜppi, R. and P. Seemann (2001). Social Capital: Securing Competitive Adventage in the New Economy, Prentice Hall.

ICOMOS (1964). "Charter of Venice." ICOMOS.

ICOMOS (1964). "INTERNATIONAL CHARTER FOR THE CONSERVATION AND RESTORATION OF MONUMENTS AND SITES."

ICOMOS (1987). "Charter for The Conservation of Historic Towns and Urban Areas."

ICOMOS (1994). Nara Document on Authenticity. Nara Conference on Authenticity.

ICOMOS (1999). The Burra Charter: The Australia ICOMOS charter for places of cultural significance 1999: with associated guidelines and code on the ethics of co-existence, Australia ICOMOS.

ICOMOS (1999). The Burra Charter: The Australia ICOMOS charter for places of cultural significance 1999: with associated guidelines and code on the ethics of co-existence, Australia ICOMOS.

ICOMOS (2008). "Quebec Declaration ON THE PRESERVATION OF THE SPIRIT OF PLACE."

ICOMOS (2010). Changing world, changing views of heritage: heritage and social change. Dublin Scientific Symposium, October 2010, Dublin, Ireland.

Issarathumnoon, W. (2004). "The Machizukuri bottom-up approach to conservation of historic communities: Lesson for Thailand." Japan: the University of Tokyo.

Jiang, S. (2012). "Analysis on the Song Marketplace Building and Space from Song Painting [从宋画中看宋朝市井街巷建筑空间]." Architectural Culture: 147-150.

Jiang, Z., B. Ma, et al. (2009). "Research on the urban redevelopmet of old houses after the Real Right Law 2007." Shanghai Legislative Research 7.

Jones, V. N. and M. Woolcock (2007). "Using mixed methods to assess social capital in low income countries: A practical guide." Brooks World Poverty Institute Working Paper Series.

Jütte, W. (2007). "Co-operation, networks and learning regions." Social capital, lifelong learning and the management of place: an international perspective: 95.

Kawachi, I. and L. Berkman (2000). "Social cohesion, social capital, and health." Social epidemiology: 174-190.

Kazmierczak, A. (2012). "Working together? Inter-organisational cooperation on climate change adaptation."

Kearns, G. and C. Philo (1993). Selling places : the city as cultural capital, past and present. Oxford England ; New York, : Pergamon Press.

Keiner, M. (2005). History, definition (s) and models of sustainable development, ETH, Eidgenössische Technische Hochschule Zürich.

Kennet, W. (1972). "Series Title: Year: 1972."

Ketelaar, J. E. (1990). Of heretics and martyrs in Meiji Japan: Buddhism and its persecution, Princeton University Press Princeton.

Kocabas, A. (2006). "Urban conservation in Istanbul: evaluation and re-conceptualisation." Habitat international 30(1): 107-126.

Kong, L. and B. S. Yeoh (1994). "Urban conservation in Singapore: a survey of state policies and popular attitudes." Urban Studies 31(2): 247-265.

Kremelberg, D. (2010). Practical Statistics: A Quick and Easy Guide to IBM (R) SPSS (R) Statistics, Stata, and Other Statistical Software, SAGE Publications.

Krishna, A. and E. Shrader (1999). "Social capital assessment tool." Social Capital Initiative Working Paper 22.

Landry, C. (2008). The creative city: A toolkit for urban innovators, Earthscan.

Landry, R., N. Amara, et al. (2002). "Does social capital determine innovation? To what extent?" Technological forecasting and social change 69(7): 681-701.

Larkham, P. J. (1996). Conservation and the city, Routledge,London.

Lee, L. J. (1977). "Tax Shelters Under the Tax Reform Act of 1976." Vill. L. Rev. 22: 223.

Lee, L. M., Y. M. Lim, et al. (2008). "Strategies for urban conservation: A case example of George Town, Penang." Habitat international 32(3): 293-304.

Lee, S. L. (1996). "Urban conservation policy and the preservation of historical and cultural heritage: The case of Singapore." Cities 13(6): 399-409.

Lees, L., T. Slater, et al. (2013). Gentrification, Routledge.

Lees, L., T. Slater, et al. (2010). The gentrification reader, Routledge.

Lefebvre, H. (2005). Critique of Everyday Life. Vol. 3: From Modernity to Modernism (Towards a Metaphilosophy of Daily Life), Verso.

Lewis, J. P. and B. Weber (1965). Building cycles and Britain's growth. London; New York, Macmillan St. Martin's Press.

Li, B. (2010). Research on the value of spatial-social complex of Linong neighborhoods in Shanghai [上海里弄街区的空间 - 社会复合体价值研究], Tongji University.

Li, H. (2007). "The collapse of Chinese ancient lane system——Case studies of Chang'an in Tang Dynasty and Dongjing in Song Dynasty [论中国古代里坊制的崩溃 —— 以唐长安与宋

东京为例]." Social Science (Chinese).

Li, J. (2010). Xi 'an Hui ethnic identity research [西安回族的民族认同研究], Minzu University of China.

Li, S. (1990). "Some characteristics of the Chinese dialect used by PingLuo Hui ethnic [平罗回族使用汉语方言的一些特点]." Journal of Ningxia University (Social Science Edition).

Li, T. (2010). "FROM CRITICAL REGIONALISM TO REFLEXIVE REGIONALISM: COMPARISON BETWEEN XINTIANDI AND TIANZIFANG." World Architecture(012): 122-127.

Li, X. (2004). "Marriages between Ethnic Groups in China." Population Research.

Li, Y. and Y. Lu (2005). "Shikumen Linong——Bugaoli [晚期石库门里弄 —— 步高里]." Shanghai Urban Planning Review(3): 35-39.

Liang, Z. (2008). "Place identity of a market street: a study of the interrelated architectural and social elements of Pei Ho Street in Sham Shui Po District." HKU Theses Online (HKUTO).

Lichfield, N. (1988). Economics in urban conservation. Cambridge England ; New York, : Cambridge University Press in association with Jerusalem Institute for Israel Studies.

Liebowitz, S. J. and S. E. Margolis (1995). "Path dependence, lock-in, and history." JL Econ. & Org. 11: 205.

Lin, C.-Y. and W.-C. Hsing (2009). "Culture-led urban regeneration and community mobilisation: The case of the Taipei Bao-an Temple area, Taiwan." Urban Studies 46(7): 1317-1342.

Lin, N. (2002). Social capital: A theory of social structure and action, Cambridge University Press, Cambridge,UK.

Liu (2008). "Defencing "Critique of Everyday Life"." cpfd.cnki.com.cn(004): 24-30.

Liu, S. (2011). The Research of Shanghai Longtang Modern Historical Changes and Cultural Values [近代上海弄堂演变及其文化价值研究], Zhejiang Normal University.

Liu, Y. (2010). "Thirty years research of 'Benevolence Righteousness Propriety Wisdom Faithfulness'." Henan Social Sciences(1): 187-190.

Logan, J. R. and Y. Bian (1993). "Inequalities in access to community resources in a Chinese city." Social Forces 72(2): 555-576.

Logan, W. and K. Reeves (2008). Places of pain and shame: dealing with'difficult heritage', Routledge.

Lomas, J. (1998). "Social capital and health: implications for public health and epidemiology." Social Science & Medicine 47(9): 1181-1188.

Lu, J. (1997). "Beijing's old and dilapidated housing renewal." Cities 14(2): 59-69.

Luo, X. and Y. Sha (2002). Xintiandi in Shanghai: the research of architectural history cultural history and development patterns in the process of old district reconstruction [上海新天地：旧区改造的建筑历史，人文历史与开发模式的研究], Southeast University Press.

LUO, Z. (2003). Luo Zhe Wen Li Shi Wen Hua Ming Cheng Yu Gu Jian Zhu Bo Hu Wen Ji China Architecture & Building Press

Lv, R. (1997). Research on the urban redevelopment of Huimin District[西安市回民区更新改造研究], Planning and Architectural Design institute of Shaanxi Muslim; Xi'an University of Architecture and Technology ; .

Lynch, J., P. Due, et al. (2000). "Social capital—Is it a good investment strategy for public health?" Journal of Epidemiology and Community Health 54(6): 404-408.

Lynch, K. (1992). The image of the city, MIT press.

Ma, J. (2008). Folk culture of Hui in Xi'an [西安回族民俗文化], San Qin Press.

Ma, J. (2012). "The analysis of Si-Fang structure in Musilm communities, Xi'an [西安穆斯林传统寺坊组织试析]." Studies in World Religions (5): 138-146.

Ma, Q. (2011). Internal and External Xi'an Hui Muslim Quarter: Islam Encounters Urban Modernization [回坊内外：城市现代化进程中的西安伊斯兰教研究]. Beijine, China Social Science Publisher.

Ma, Q. and Z. Ma (2010). "From Seven Mosques-Thirteen Fangs to Twelve Mosques- the study of Si-Fang structure of Hui communities in Xi'an [从"七寺十三坊"到回坊十二寺——西安回族寺坊结构变化研究之一]." The northwest national review.

Ma, S. (1981). "The spread, development, and evolution of Islam in Shaanxi [伊斯兰教在陕西的传播发展与演变]." The collection of Islam in China in Qing Dynasty.

Madgin, R. (2009). Heritage, Culture and Conservation: Managing the Urban Renaissance, VDM Publishing.

Mageean, A. (1999). "Assessing the impact of urban conservation policy and practice: the Chester experience 1955-96." Planning Perspectives 14(1): 69-97.

Magnusson, L. and J. Ottosson (2009). The evolution of path dependence, Edward Elgar Publishing, Cheltenham,UK.

Mason, R. (2002). "Assessing values in conservation planning: methodological issues and choices." Assessing the values of cultural heritage: 5-30.

Mason, R. (2005). "Economics and historic preservation." Washington, DC: The Brookings Institution.

Mason, R. (2008). Management for Cultural Landscape Preservation: Insights from Australia, na.

Miller, R. and C. Acton (2009). SPSS for social scientists, Palgrave Macmillan.

Mo, T. and D. Lu (2000). "Regeneration of Unban Form of Shanghai Lilong-Conservative Development of Xintiandi [J]." Time+ Architecture 3.

Mou, Y. and H. Tan (2008). "Cultural heritege conservation and local economic development: case study of Chongqing model [文化遗产保护与地方经济发展——以"重庆模式"为例]." Qiusuo [求索]: 20-23.

Murray, C. (2001). Making sense of place: new approaches to place marketing, Comedia Publications.

Nahapiet, J. and S. Ghoshal (1998). "Social capital, intellectual capital, and the organizational advantage." Academy of management review 23(2): 242-266.

Nankervis, M. (1988). "Conservation controls in Melbourne." Cities 5(2): 137-143.

Narayan, D. and M. F. Cassidy (2001). "A dimensional approach to measuring social capital: development and validation of a social capital inventory." Current Sociology 49(2): 59-102.

Nasser, N. (2003). "Planning for urban heritage places: reconciling conservation, tourism, and sustainable development." Journal of planning literature 17(4): 467.

Nichols, R. M., Ed. (1996). Main street approach: urban conservation and economic development, Center for Built Environment.

NPCSC (2002). Law of the people's Republic of China on Protection of Clutural Relics [中华人民共和国文物保护法]. The National People's Congress standing committee of the People's Republic of China [中华人民共和国全国人民代表大会常务委员会公报].

Nyseth, T. and J. Sognnæs (2013). "Preservation of old towns in Norway: Heritage discourses, community processes and the new cultural economy." Cities 31: 69-75.

Orbasli, A. (2000). Tourists in historic towns: Urban conservation and heritage management, Taylor & Francis.

Ostrom, E. (2000). "Social Capital: Popular Fanatic or Basic Conception." Social Capital, World Bank.

Owen, C. (1990). "Tourism and urban regeneration." Cities 7(3): 194-201.

Ozmete, E. (2011). "UNDERSTANDING OF SOCIAL CAPITAL WITH ECOSYSTEM APPROACH." INTERNATIONAL JOURNAL Of ACADEMIC RESEARCH Vol. 3. No. 3.: 333-339.

Peel, D. (2003). "Town Centre Management: multi-stakeholder evaluation. Increasing the sensitivity of paradigm choice." Planning Theory & Practice 4(2): 147-164.

Pena, M. V. J. and H. Lindo-Fuentes (1998). "Community Organization, Values and Social Capital in Panama. ." Central America Country Management Unit Economic Notes No. 9,(The World Bank, Washington, D.C.).

Pena, M. V. J., H. Lindo-Fuentes, et al. (1998). Community Organization, Values and Social Capital in Panama, Central America Country Management Unit, Latin America and Caribbean Region, the World Bank.

Pendlebury, J. (2008). Conservation in the Age of Consensus, Routledge.

Piao, C. (2003). "The Theoretical Explanation and Modern Content of the Structure of Grade [" 差序格局 " 的理论诠释及现代内涵]." Sociolgical Research 1: 002.

Pickard, R. (2001). Management of historic centres, Taylor & Francis, London/New York.

Ping, H. O. S. H. U. (2007). "Partners in conservation-Communities, Contestation and Conflict in Komodo National Park, Indonesia."

Portes, A. (2000). "Social capital: Its origins and applications in modern sociology." Knowledge and social capital: foundations and applications: 43-67.

Portes, A. and P. Landolt (1996). "The downside of social capital." The american prospect 26(94): 18-21.

Pretty, J. (2003). "Social capital and the collective management of resources." Science 302(5652): 1912-1914.

Punter, J. (1987). "A history of aesthetic control: Part 2, 1953-1985: The control of the external appearance development in England and Wales." Town Planning Review 58(1): 29.

Putnam, R. (2000). "Bowling alone: The collapse and revival of American community." Nova Iorque, Simon Schuster.

Putnam, R. D. (1993). "The prosperous community." The american prospect 4(13): 35-42.

Putnam, R. D. (1995). "Bowling alone: America's declining social capital." Journal of democracy 6(1): 65-78.

Putnam, R. D. (2002). Democracies in flux: The evolution of social capital in contemporary society, Oxford University Press, USA.

Putnam, R. D., R. Leonardi, et al. (1993). Making democracy work : civic traditions in modern Italy. Princeton, N.J., : Princeton University Press.

Putnam, R. D., R. Leonardi, et al. (1994). Making democracy work: Civic traditions in modern Italy, Princeton Univ Pr.

Qiu, S. (2001). "Descriptions of the Bureau of Hadji in the Documents Excavated in Heicheng [从黑城出土文书看元" 回回哈的司"]." Journal of Nanjing University (Philosophy, Humanities and Social Sciences): 152-160.

Rapoport, A. (1990). History and precedent in environmental design, Plenum Press New York.

Rapoport, A. and J. Silverberg (1973). "The Mutual Interaction of People and Their Built Environment." Man-Environment Systems 3(4): 235-236.

Roberts, P. (2000). "The evolution, definition and purpose of urban regeneration." Urban regeneration: a handbook: 9-36.

Robertson, K. A. (2004). "The main street approach to downtown development: An examination of the four-point program." Journal of Architectural and Planning Research 21(1): 55-73.

Rodwell, D. (2007). Conservation and sustainability in historic cities, Wiley-Blackwell,Oxford.

Roodhouse, S. (2010). Cultural Quarters: Principles and Practice, Intellect Ltd.

Ruan, Y. and L. Lin (2003). "Authenticity in Relation to the Conservation of Cultural Heritage [J]." Journal of Tongji University Social Science Section 2.

Ruan, Y. and M. Sun (2001). "The study on some issues related to the conservation and planning for the historic streets and areas in China [我国历史街区保护与规划的若干问题研究]." Shanghai Urban Planning Review: 10(25): 25-32.

RUAN, Y. and F. Yuan (2008). Conservation and Development of the Historic Water Towns in the South of Yangtze River [J].

Ruan, Y. and S. Zhang (2004). The Promotion of Industry Heritage Conservation on Cultural Business. URBAN PLANNING FORUM.

Ruan, Y., Y. Zhang, et al. (2003). "Conservation of City Heritage in The Market Economy." Urban Planning Review 12: 48-51.

Ruskin, J. (1884). The Seven Lamps of Architecture (1849); reprint, New York, John Wiley & Sons.

Salazar, M. P. (2006). Promises and challenges of urban community-oriented conservation: The case of greenways in Detroit (Michigan), Michigan State University.

Sandercock, L. (1998). "Framing insurgent historiographies for planning." Making the invisible visible: A multicultural planning history: 1-33.

Sandercock, L. and P. Lysiottis (1998). "Towards cosmopolis: Planning for multicultural cities."

Sanoff, H. (2000). Community participation methods in design and planning, John Wiley & Sons Inc.

Sauer, C. O. (1925). "The Morphology of Landscape " Geography 2: 19-54.

Savage, V. R., S. Huang, et al. (2004). "The Singapore River thematic zone: sustainable tourism in an urban context." The Geographical Journal 170(3): 212-225.

Schienstock, G. (2011). "Path dependency and path creation: continuity vs fundamental change in national economies." Journal of Futures Studies 15(4): 63-76.

Scott, J. (1991). Social network analysis : a handbook. London, Sage Publications.

Scott, J. (1991). Social network analysis: A handbook, Sage Publications (London and Newbury Park, Calif.).

Scott, J. (2000). Social network analysis : a handbook. London, Sage Publications.

Shan, Q. (2007). "The protection of urban cultural heritage and urban construction [城市文化遗产保护与文化城市建设]." Citiy planning review: 9-23.

Shao, Y., L. Zhang, et al. (2004). "Conservation and social development of Ancient Town of Lijiang [Shijie wenhua yichan lijiang gucheng de baohu he shehui fazhan]." Ideal Space 6: 52-55.

Sharif Shams, I. (2006). Sustainable urban conservation : the role of public participation in the conservation of urban heritage in old Dhaka. Ph.D.

Shin, H. B. (2010). "Urban conservation and revalorisation of dilapidated historic quarters: The case of Nanluoguxiang in Beijing." Cities 27: S43-S54.

Shinohara, H. (2009). Mutation of Tianzifang, Taikang Road, Shanghai. 4th International Conference of the International Forum on Urbanism,"The New Urban Question: Urbanism beyond Neo-Liberalism," Amsterdam/Delft.

Siegenthaler, P. (2003). Creation Myths for the Preservation of Tsumago Post-town.

Sim, D. (1982). Change in the city centre, Gower.

Slater, T. R. (1984). "Preservation, conservation and planning in historic towns." Geographical Journal: 322-334.

Smith, L. (2006). Uses of heritage, Routledge.

Smith, N. (1979). "Toward a theory of gentrification a back to the city movement by capital, not people." Journal of the American Planning Association 45(4): 538-548.

Snyder, M. R. (2008). "The role of heritage conservation districts in achieving community improvement."

Stiglitz, J. E. (2000). "Formal and informal institutions." Social capital: A multifaceted perspective: 59-70.

Strange, I. (1997). "Planning for change, conserving the past: towards sustainable development policy in historic cities?" Cities 14(4): 227-233.

Su, X. (2010). "Urban conservation in Lijiang, China: Power structure and funding systems." Cities 27(3): 164-171.

Tan, C. and C. S. Tan (2014). "Fostering Social Cohesion and Cultural Sustainability: Character and Citizenship Education in Singapore." Diaspora, Indigenous, and Minority Education 8(4): 191-206.

Tan, L. (2006). Revolutionary Spaces in Globalization: Beijing's Dashanzi Arts District'.

Tarn, J. N. (1985). "Urban regeneration: the conservation dimension." Town Planning Review 56(2): 245.

Taschereau, D. M. (2001). Urban social sustainability: opportunities for Southeast False Creek, Simon Fraser University.

Teo, P. and S. Huang (1995). "Tourism and heritage conservation in Singapore." Annals of Tourism Research 22(3): 589-615.

Theobald, W. F. (1998). Global tourism, Butterworth-Heinemann.

Tianwei, M. and L. Di (2000). "Regeneration of Unban Form of Shanghai Lilong-Conservative Development of Xintiandi [J]." TIME+ ARCHITECTURE 3.

Tiesdell, S., T. Oc, et al. (1996). Revitalizing historic urban quarters, Architectural Press,Oxford.

Tweed, C. and M. Sutherland (2007). "Built cultural heritage and sustainable urban development." Landscape and Urban Planning 83(1): 62-69.

UNESCO (1967). "Civic Amenities Act ".

UNESCO (1972). "Convention Concerning the Protection of the World Cultural and Natural Heritage."

UNESCO (1976). "Recommendation concerning the Safeguarding and Contemporary Role of Historic Areas."

UNESCO (2003). "Convention for the Safeguarding of the Intangible Cultural Heritage." from http://www.unesco.org/culture/ich/index.php?lg=en&pg=00006.

UNESCO (2011). Proposals Concerning The Desirability of A Standard-Setting Instrument on Historic Urban Landscapes. General Coference, 36th Session. Paris.

URA (2011). Conservation Guidelines for Historic Districts. S. Urban Redevelopment Authority.

Urry, J. (1995). Consuming places, Psychology Press.

Urry, J. (1995). "How societies remember the past." The Sociological Review 43(S1): 45-65.

Uysal, Ü. E. (2011). "An urban social movement challenging urban regeneration: The case of Sulukule, Istanbul." Cities.

Uzzell, D., E. Pol, et al. (2002). "Place identification, social cohesion, and enviornmental sustainability." Environment and Behavior 34(1): 26-53.

Uzzi, B. (1997). "Social structure and competition in interfirm networks: The paradox of embeddedness." Administrative science quarterly: 35-67.

Wai, A. W. T. (2006). "Place promotion and iconography in Shanghai's Xintiandi." Habitat international 30(2): 245-260.

Wakefield, S. E. L. and B. Poland (2005). "Family, friend or foe? Critical reflections on the relevance and role of social capital in health promotion and community development." Social Science & Medicine 60(12): 2819-2832.

Wallin, E. (2007). "Social Capital, Lifelong Learning and the Management of Place." Social capital, lifelong learning and the management of place: an international perspective: 161-180.

Wang, B. and Z. Zhang (2012). "Case study of Shanghai Xintiandi design [上海新天地规划设计案例研究]." Shanxi Architecture 38(19): 26-27.

Wang, D. (2002). "Religious system and Islamic laws of Hui ethnic in Yuan Dynasty [元代回回人的宗教制度与伊斯兰教法]." The Hui studies: 44-50.

Wang, J. (2009). "'Art in capital': Shaping distinctiveness in a culture-led urban regeneration project in Red Town, Shanghai." Cities 26(6): 318-330.

Wang, J., Y. Ruan, et al. (1999). Historical and cultural city protection theory and planning [历史文化名城保护理论与规划]. Shanghai, Tongji university press.

Wang, J., Z. Yao, et al. (2009). "Preservation and Regeneration Via Hai Pai Cultural Renaissance–A Case Study of Tianzifang Creative Quarter in Shanghai."

Wang, L. and J. Wang (1998). "The research on planning methods for urban historic area conservation [历史街区保护规划编制方法研究]." City Planning Review: 37-39.

Wang, Q. and T. Ma (1982). "Religious sect and school of Islam in China [中国伊斯兰教的教派和门宦]." Journal of northwest university for nationalities (Social sciences).

Wang, S. Y. (2008). Tradition, memory and the culture of place: Continuity and change in the ancient city of Pingyao, China, ProQuest.

Wang, W. (2004). "Public participation in urban heritage conservation [历史街区的保护和公众参与]." Journal of Nanjing University of Technology (Social sciences): 58-61.

Wang, Y. (1993). "The appearing and early forms of Jing Tang in Hui ethnic [回族经堂教育的产生及早期形态]." The Hui studies.

Wang, Y. (1997). "Zhao-Wu-Jiu-Xing Hu in Hou-Tang, Hou-Jin, and Hou-Han Dynasties [后唐、后晋、后汉王朝的昭武九姓胡]." North West Ethno-national Studies.

WCED (1987). "Report of the World Commission on Environment and Development: Our Common Future ".

Wells, J. C. (2007). "The plurality of truth in culture, context, and heritage: a (mostly) post-structuralist analysis of urban conservation charters."

Wells, J. C. (2010). "Our history is not false: perspectives from the revitalisation culture." International Journal of Heritage Studies 16(6): 464-485.

Wells, J. C. and E. D. Baldwin (2012). "Historic preservation, significance, and age value: A comparative phenomenology of historic Charleston and the nearby new-urbanist community of I'On." Journal of Environmental Psychology 32(4): 384-400.

Whitehand, J. (1990). "Townscape management: ideal and reality." The built form of Western cities: 370-393.

Whitehand, J., K. Gu, et al. (2011). "Urban morphology and conservation in China." Cities.

Whitehand, J. W. R. (1992). "Recent advances in urban morphology." Urban studies 29(3-4): 619.

Whitehand, J. W. R., K. Gu, et al. (2011). "Urban morphology and conservation in China." Cities 28(2): 171-185.

Whyte, W. H. and M. A. S. o. N. York (1980). The social life of small urban spaces, Conservation Foundation Washington, DC.

Wideman, M. (1998). A community development approach to heritage tourism in small towns: a case study of Millbrook, Ontario.

Widodo, J. (2004). The Boat and the city: Chinese diaspora and the Architecture of Southeast Asian coastal cities, Chinese Heritage Centre.

Wing, H. C. and S. L. Lee (1980). "The characteristics and locational patterns of wholesale and service trades in the central area of Singapore." Singapore Journal of Tropical Geography 1(1): 23-36.

Wong, J. and J. Rong (2001). "Conservation and Adaptive Reuse of Historical Industrial Buildings and Sites [J]." TIME+ ARCHITECTURE 4.

Wong, S. (2008). Exploring 'unseen' social capital in community participation: everyday lives of poor mainland Chinese migrants in Hong Kong, Amsterdam University Press.

Wong, Y. and T. K. Leung (2001). Guanxi: Relationship marketing in a Chinese context, International Business Press New York, NY.

Woolcock, M. (1998). "Social capital and economic development: Toward a theoretical synthesis and policy framework." Theory and society 27(2): 151-208.

Woolcock, M. (2001). "The place of social capital in understanding social and economic outcomes." Canadian Journal of Policy Research 2(1).

Works, U. S. C. H. C. o. P., Transportation, et al. (1976). Public Buildings Cooperative Use Act of 1976: Report to Accompany HR 15134, US Govt. Print. Off.

Worthington, J., J. Warren, et al. (1998). Context: New Buildings in Historic Settings, Oxford: Architectural Press.

Wu, F. (1985). "Markets and business in Tang Chang'an [唐长安的市场和商业]." Journal of Northwest University (Social Sciences).

Wu, F. (2000). "The global and local dimensions of place-making: remaking Shanghai as a world city." Urban Studies 37(8): 1359-1377.

Wu, H. (1999). "The form and the layout characteristics of Chang 'an in tang dynasty and Wudai dynasties [论唐末五代长安城的形制和布局特点]." Journal of Chinese Historic Geography.

Wu, L. (1991). "Rehabilitation in Beijing." Habitat international 15(3): 51-66.

XCPB (1980). Xi'an Master Plan 1980-2000. Xi'an.

XCPB (2004). Xi'an Master Plan 2004-2020. Xi'an.

Xiao, J. (2007). "Make use of the cultural creative industry to activate urban heritage——case study of M50 in Moganshan Road [利用文化创意产业激活城市遗产 —— 以 M50 为代表的莫干山路历史工厂区为例]." Shanghai real estate(2): 47-50.

Xie, C. (2013). "Four erroneous tendancies in current conservation of cultural relices [当前文物工作应当防止的四种错误倾向 —— 在全国人大常委会召开的纪念文物保护法颁布 30 周年座谈会上的发言]." City Planning Review.

Xu, F., H. Han, et al. (2006). "The Conceptive Conservation Plan of the Historical Industrial Architecture along the Moganshan Road on the South Bank of the Suzhou Creek [苏州河南岸莫干山路地块历史产业建筑群概念性保护规划]." Journal of shanghai institute of technology

5(4): 307-312.

Xu, H. and W. Tao (2001). "Managing side effects of cultural tourism development: The Case of Zhouzhuang." Chinese Geographical Science 11(4): 356-365.

Xu, M. (2004). Urban context: new development mothods for old communities in Shanghai [城市的文脉：上海中心城旧住区发展方式新论], Xuelin Publishing House.

Xu, Q. (2007). "Public Participation and the Sustainable Urban Renewal:Inspiration from the Practice of old City Renewal in Yangzhou [公众参与和可持续的老城更新——扬州老城更新的实践与启示]." Modern Urban Research: 4-9.

Xue, D.-q., S.-m. Yao, et al. (2000). "Contacting on the function and optimizing on the structure in Guanzhong urban agglomeration." ECONOMIC GEOGRAPHY.

Yan, Y. (2006). ""Chaxu geju" and the Notion of Hierarchy in Chinese Culture [差序格局与中国文化的等级观]." Sociolgical Studies 4: 201-213.

Yang, W. (2006). "Hui Communities under the transformation of relation between nation and society in Ming and Qing Dynasties——Case study of Xi'an Hui Communites [明清时期国家与社会关系转型境遇下的回族社区——以历史上西安回族社区文化变迁为视点]." Journal of Heilongjiang Minzu: 89-97.

Yang, Y. (2013). "How to realize social solidarity in the society under the pattern of CHAXUGEJU ["差序格局"社会中团结何以可能]." Young Society(11): 321-321.

Yang, Y. R. and C. Chang (2007). "An urban regeneration regime in China: A case study of urban redevelopment in Shanghai's Taipingqiao area." Urban Studies 44(9): 1809-1826.

Yao, J. (2008). The Regeneration of Atypical Historic Districts. Shanghai, Tongji University.

Yao, P. and Y. Zhao (2009). "How to Protect and Utilize Historical Heritage: a Case Study with Xintiandi Shopping Mall in Shanghai." Journal of Eastern Liaoning University (Natural Science) 1.

Yao, P. and Y. Zhao (2009). "The issues of the renew and protection of Shanghai Xintiandi [基于上海新天地对历史遗产保护利用问题的思考]." Journal of Eastern Liaoning

University(Nature Science) 16(1): 75-78.

Yeoh, B. S. A. and S. Huang (1996). "The conservation-redevelopment dilemma in Singapore:: The case of the Kampong Glam historic district." Cities 13(6): 411-422.

Yiftachel, O. and D. Hedgcock (1993). "Urban social sustainability: The planning of an Australian city." Cities 10(2): 139-157.

Yin, R. K. (1984). Case study research: design and methodology, California: SAGE Publications. Liite.

Yip, M. T. L. and URA (2008). "Singapore's experience in conservation."

Young, I. M. (1990). Justice and the Politics of Difference, Princeton University Press,Princeton.

Yu, Z. (2010). The research on the development and protection and Renewal of "Lilong" neighborhood in Shanghai [上海里弄住宅的演变及保护和更新研究], China Central Academy of Fine Arts.

Yu-fang, Y. (2011). "Problems and Solutions of Zhouzhuang Becoming Exquisite Tourism Product." Journal of Jiangsu Vocational and Technical Institute of Economics and Commerce.

Zhang, C., J. WANG, et al. (2008). "The impact of local creative milieu and space on cultural and creative activities in cities: The case of Nanluoguxiang in Beijing [J]." Geographical Research 2.

Zhang, J. (2009). Lao Zheng and his Tianzi Fang. Blog.Sina. Shanghai.

Zhang, S. (1999). "The objective and method of conservation of historic towns." Urban Planning(007): 50-53.

Zhang, s. (2006). "Conservation Strategy Of Urban Heritage In Shanghai [上海城市遗产的保护策略]." City Planning Review 30(2): 49-54.

Zhang, S. and P. Chen (2010). "The Conservation of Industrial Heritage and the Development of Creative Industrial " Architectural Journal(012): 12-16.

Zhang, S., M. Lu, et al. (2007). "Does the Strength of Social Capital on Poverty Reduction Fall or Rise during Marketization?-Evidence from Rural China." CHINA ECONOMIC QUARTERLY-BEIJING- 6(2): 539.

Zhang, Y. (2008). "Xi'an Muslim Quarter: opportunities and challenges for public participation in historic conservation." HKU Theses Online (HKUTO).

Zhao, M., J. Bao, et al. (1998). The reform of land use system and the urban and rural developments [土地使用制度改革与城乡发展], Tongji University Press.

Zheng, Y. (2006). A Urban Morphology Study on Quanzhou in Ming&Qing Dynasty, HuaQiao Univiersity,China.

ZHOU, S. and X. SHEN (2008). "Local Cultural Embeddedness of the Fine Arts Industry in Beijing [J]." Human Geography 2.

Zhu, D. and X. Zhu (2010). "Two lanes in Shanghai-Textual Study on the Relation between Bugao Lane and Jianye Lane [两条上海里弄——步高里和建业里的关系考证]." Residential science and technology 30(12): 38-42.

Zhu, J., L.-L. Sim, et al. (2006). "Place-remaking under property rights regimes: a case study of Niucheshui, Singapore." Institute of Urban & Regional Development.

Zhu, R. (2009). "Tianzi Fang--the exploration of urban sustainable development model in China [解读田子坊——我国城市可持续发展模式的探索]."

Zukin, S. (2009). "Changing landscapes of power: opulence and the urge for authenticity." International Journal of Urban and Regional Research 33(2): 543-553.

Appendix Principle Component Analysis (PCA)

Five indicators of cognitive social cohesion are simplified through the extraction method of Principle Component Analysis (PCA).

Hui Fang, Xi'an

Factor analysis of social cohesion indicator——Benevolence (仁) in Hui Fang   Table 1

Total Variance Explained

Component	Initial Eigenvalues			Extraction Sums of Squared Loadings	
	Total	% of Variance	Cumulative %	Total	% of Variance
1	4.140	68.994	68.994	4.140	68.994
2	.599	9.991	78.986		
3	.505	8.424	87.410		
4	.349	5.817	93.227		
5	.258	4.304	97.530		
6	.148	2.470	100.000		

Component	Extraction Sums of Squared Loadings
	Cumulative %
1	68.994
2	
3	
4	
5	
6	

Component Matrix

	Component
	1
R1	.896
R2	-.775
R3	.900
R4	.762
R5	.888
R6	.746

Factor analysis of social cohesion indicator——Righteousness（义）in Hui Fang   Table 2

Total Variance Explained

Component	Initial Eigenvalues			Extraction Sums of Squared Loadings	
	Total	% of Variance	Cumulative %	Total	% of Variance
1	3.410	56.832	56.832	3.410	56.832
2	1.112	18.530	75.362	1.112	18.530
3	.509	8.488	83.850		
4	.435	7.257	91.106		
5	.359	5.981	97.087		
6	.175	2.913	100.000		

Component	Extraction Sums of Squared Loadings	Rotation Sums of Squared Loadings		
	Cumulative %	Total	% of Variance	Cumulative %
1	56.832	2.775	46.246	46.246
2	75.362	1.747	29.116	75.362
3				
4				
5				
6				

Component Matrix

	Component	
	1	2
Y1	.789	
Y2	.615	.609
Y3	.877	
Y4	.845	
Y5	-.455	.788
Y6	.847	

Factor analysis of social cohesion indicator——Propriety（礼）in Hui Fang   Table 3

Total Variance Explained

Component	Initial Eigenvalues			Extraction Sums of Squared Loadings	
	Total	% of Variance	Cumulative %	Total	% of Variance
1	4.371	72.852	72.852	4.371	72.852
2	.753	12.548	85.400		
3	.352	5.861	91.261		
4	.262	4.375	95.636		

5	.166	2.773	98.409		
6	.095	1.591	100.000		
Component	Extraction Sums of Squared Loadings				
	Cumulative %				
1					72.852
2					
3					
4					
5					
6					

Component Matrix

	Component
	1
L1	.882
L2	.906
L3	.907
L4	.940
L5	-.865
L6	.561

Factor analysis of social cohesion indicator——Wisdom（智）in Hui Fang　　Table 4

Total Variance Explained

Component	Initial Eigenvalues			Extraction Sums of Squared Loadings	
	Total	% of Variance	Cumulative %	Total	% of Variance
1	3.141	52.357	52.357	3.141	52.357
2	.830	13.831	66.188		
3	.774	12.901	79.088		
4	.524	8.737	87.825		
5	.408	6.804	94.628		
6	.322	5.372	100.000		
Component	Extraction Sums of Squared Loadings				
	Cumulative %				
1					52.357
2					
3					
4					
5					
6					

Component Matrix	
	Component
	1
Z1	.592
Z2	.818
Z3	.777
Z4	.728
Z5	.786
Z6	.608

Factor analysis of social cohesion indicator——Faithfulness(信)in Hui Fang   Table 5

Total Variance Explained					
Component	Initial Eigenvalues			Extraction Sums of Squared Loadings	
	Total	% of Variance	Cumulative %	Total	% of Variance
1	4.201	70.010	70.010	4.201	70.010
2	.655	10.917	80.927		
3	.409	6.812	87.739		
4	.278	4.641	92.380		
5	.254	4.227	96.607		
6	.204	3.393	100.000		

	Extraction Sums of Squared Loadings
Component	Cumulative %
1	70.010
2	
3	
4	
5	
6	

Component Matrix	
	Component
	1
X1	.882
X2	.892
X3	.826
X4	.768
X5	-.844
X6	.801

Tianzi Fang, Shanghai

Factor analysis of social cohesion indicator——Benevolence (仁) in Tianzi Fang   Table 6

Component	Initial Eigenvalues			Extraction Sums of Squared Loadings	
	Total	% of Variance	Cumulative %	Total	% of Variance
1	4.064	67.741	67.741	4.064	67.741
2	.945	15.756	83.497		
3	.444	7.401	90.898		
4	.323	5.377	96.274		
5	.157	2.614	98.888		
6	.067	1.112	100.000		

Total Variance Explained

Component	Extraction Sums of Squared Loadings
	Cumulative %
1	67.741
2	
3	
4	
5	
6	

Component Matrix

	Component
	1
R1	.784
R2	-.652
R3	.865
R4	.923
R5	.870
R6	.817

Factor analysis of social cohesion indicator——Righteousness (义) in Tianzi Fang   Table 7

Total Variance Explained

Component	Initial Eigenvalues			Extraction Sums of Squared Loadings	
	Total	% of Variance	Cumulative %	Total	% of Variance
1	4.144	69.074	69.074	4.144	69.074
2	.706	11.762	80.837		
3	.435	7.242	88.079		

4	.386	6.431	94.509		
5	.199	3.312	97.821		
6	.131	2.179	100.000		
Component	Extraction Sums of Squared Loadings				
	Cumulative %				
1					69.074
2					
3					
4					
5					
6					

Component Matrix

	Component
	1
Y1	.823
Y2	.840
Y3	.939
Y4	.851
Y5	-.879
Y6	.618

Factor analysis of social cohesion indicator——Propriety (礼) in Tianzi Fang   Table 8

Total Variance Explained

Component	Initial Eigenvalues			Extraction Sums of Squared Loadings	
	Total	% of Variance	Cumulative %	Total	% of Variance
1	4.175	69.577	69.577	4.175	69.577
2	.665	11.083	80.660		
3	.387	6.456	87.116		
4	.357	5.949	93.065		
5	.225	3.744	96.810		
6	.191	3.190	100.000		
Component	Extraction Sums of Squared Loadings				
	Cumulative %				

1		69.577
2		
3		
4		
5		
6		

Component Matrix

	Component
	1
L1	.714
L2	.862
L3	.872
L4	.896
L5	-.808
L6	.840

Factor analysis of social cohesion indicator——Wisdom（智）in Tianzi Fang   Table 9

Total Variance Explained

Component	Initial Eigenvalues			Extraction Sums of Squared Loadings	
	Total	% of Variance	Cumulative %	Total	% of Variance
1	4.360	72.672	72.672	4.360	72.672
2	.489	8.145	80.817		
3	.346	5.774	86.591		
4	.324	5.403	91.993		
5	.278	4.626	96.620		
6	.203	3.380	100.000		

Component	Extraction Sums of Squared Loadings
	Cumulative %
1	72.672
2	
3	
4	
5	
6	

Component Matrix

	Component
	1
Z1	.849

Z2	.868
Z3	.894
Z4	.843
Z5	.854
Z6	.804

Factor analysis of social cohesion indicator——Faithfulness（信）in Tianzi Fang　　Table 10

Total Variance Explained					
Component	Initial Eigenvalues			Extraction Sums of Squared Loadings	
	Total	% of Variance	Cumulative %	Total	% of Variance
1	4.253	70.885	70.885	4.253	70.885
2	.757	12.619	83.504		
3	.396	6.603	90.107		
4	.267	4.443	94.550		
5	.203	3.382	97.932		
6	.124	2.068	100.000		

Component	Extraction Sums of Squared Loadings
	Cumulative %
1	70.885
2	
3	
4	
5	
6	

Component Matrix	
	Component
	1
X1	.859
X2	.888
X3	.881
X4	.935
X5	-.868
X6	.569